T0205401

Advances in Intelligent Systems and Computing

Volume 892

Series editor

Janusz Kacprzyk, Polish Academy of Sciences, Warsaw, Poland
e-mail: kacprzyk@ibspan.waw.pl

The series "Advances in Intelligent Systems and Computing" contains publications on theory, applications, and design methods of Intelligent Systems and Intelligent Computing. Virtually all disciplines such as engineering, natural sciences, computer and information science, ICT, economics, business, e-commerce, environment, healthcare, life science are covered. The list of topics spans all the areas of modern intelligent systems and computing such as: computational intelligence, soft computing including neural networks, fuzzy systems, evolutionary computing and the fusion of these paradigms, social intelligence, ambient intelligence, computational neuroscience, artificial life, virtual worlds and society, cognitive science and systems, Perception and Vision, DNA and immune based systems, self-organizing and adaptive systems, e-Learning and teaching, human-centered and human-centric computing, recommender systems, intelligent control, robotics and mechatronics including human-machine teaming, knowledge-based paradigms, learning paradigms, machine ethics, intelligent data analysis, knowledge management, intelligent agents, intelligent decision making and support, intelligent network security, trust management, interactive entertainment, Web intelligence and multimedia.

The publications within "Advances in Intelligent Systems and Computing" are primarily proceedings of important conferences, symposia and congresses. They cover significant recent developments in the field, both of a foundational and applicable character. An important characteristic feature of the series is the short publication time and world-wide distribution. This permits a rapid and broad dissemination of research results.

More information about this series at http://www.springer.com/series/11156

Michał Choraś · Ryszard S. Choraś
Editors

Image Processing and Communications Challenges 10

10th International Conference, IP&C'2018
Bydgoszcz, Poland, November 2018,
Proceedings

Editors
Michał Choraś
Institute of Telecommunications
and Computer Science
University of Science and Technology, UTP
Bydgoszcz, Poland

Ryszard S. Choraś
Institute of Telecommunications
and Computer Science
University of Science and Technology, UTP
Bydgoszcz, Poland

ISSN 2194-5357   ISSN 2194-5365   (electronic)
Advances in Intelligent Systems and Computing
ISBN 978-3-030-03657-7   ISBN 978-3-030-03658-4   (eBook)
https://doi.org/10.1007/978-3-030-03658-4

Library of Congress Control Number: 2018960422

This Springer imprint is published by the registered company Springer Nature Switzerland AG
The registered company address is: Gewerbestrasse 11, 6330 Cham, Switzerland

Preface

This volume of AISC contains the proceedings of the International Conference on Image Processing and Communications, IP&C 2018, held in Bydgoszcz, November 2018.

The International Conference on Image Processing and Communications started ten years ago as a forum for researchers and practitioners in the broad fields of image processing, telecommunications, and pattern recognition. It is known for its scientific level, the special atmosphere, and attention to young researchers. As in the previous years, IP&C 2018 was organized by the UTP University of Science and Technology and was hosted by the Institute of Telecommunications and Computer Sciences of the UTP University.

IP&C 2018 brought together the researchers, developers, practitioners, and educators in the field of image processing and computer networks. IP&C has been a major forum for scholars and practitioners on the latest challenges and developments in IP&C.

The conference proceedings contains high-level papers which were selected through a strict review process. The presented papers cover all aspects of image processing (from topics concerning low-level to high-level image processing), pattern recognition, novel methods and algorithms as well as modern communications.

Without the high-quality submissions from the authors, the success of the conference would not be possible. Therefore, we would like to thank all authors and also the reviewers for the effort they put into their submissions and evaluation.

The organization of such an event is not possible without the effort and the enthusiasm of the people involved. The success of the conference would not be possible without the hard work of the local Organizing Committee.

Last, but not least, we are grateful to dr Karolina Skowron for her management work and to dr Adam Marchewka for hard work as the Publication Chair, and also to Springer for publishing the IP&C 2018 proceedings in their Advances in Intelligent Systems and Computing series.

Michał Choraś
Conference Chair

Organization

Organization Committee

Conference Chair

Michał Choraś, Poland

Honorary Chairs

Ryszard Tadeusiewicz, Poland
Ryszard S. Choraś, Poland

International Program Committee

Kevin W. Bowyer, USA
Dumitru Dan Burdescu, Romania
Christophe Charrier, France
Leszek Chmielewski, Poland
Michał Choraś, Poland
Andrzej Dabrowski, Poland
Andrzej Dobrogowski, Poland
Marek Domański, Poland
Kalman Fazekas, Hungary
Ewa Grabska, Poland
Andrzej Kasiński, Poland
Andrzej Kasprzak, Poland

Marek Kurzyński, Poland
Witold Malina, Poland
Andrzej Materka, Poland
Wojciech Mokrzycki, Poland
Sławomir Nikiel, Poland
Zdzisław Papir, Poland
Jens M. Pedersen, Denmark
Jerzy Pejaś, Poland
Leszek Rutkowski, Poland
Khalid Saeed, Poland
Abdel-Badeeh M. Salem, Egypt

Organizing Committee

Łukasz Apiecionek
Sławomir Bujnowski
Piotr Kiedrowski
Rafał Kozik
Damian Ledziński
Zbigniew Lutowski
Adam Marchewka
   (Publication Chair)

Beata Marciniak
Tomasz Marciniak
Ireneusz Olszewski
Karolina Skowron
   (Conference Secretary)
Mścisław Śrutek
Łukasz Zabłudowski

Contents

Vision Processing

Ten Years of Image Processing and Communications

Ryszard S. Choraś, Agata Giełczyk[✉], and Michał Choraś

Faculty of Telecommunications, Computer Science and Electrical Engineering,
UTP University of Science and Technology, Bydgoszcz, Poland
agata.gielczyk@utp.edu.pl

Abstract. Image processing and communications have become emerging domains for researchers and societies all over the world. Both are widely implemented and become reality in everyday matters. In this article authors present an overview of the trends and reflect on aspects discussed during the Image Processing and Communications Conferences taking place in Bydgoszcz, Poland during last 10 years. The paper aims to reflect on this great event and its scientific contents.

Keywords: Image processing · Biometrics · Pattern recognition
Applications · Video processing · Image quality · Network
Cyber security · Cloud computing

1 Introduction

In the mid-90s of the last century, prof. Ryszard S. Choras has created the Image Processing Group research group focusing on signal and image processing and also started to publish in English the international academic quarterly journal Image Processing & Communications as its Editor-in-Chief. In the early years of this group activity the research tasks concerned image compression using various types of transformations (e.g. Fourier, Hadamard, DCT, Slant, DWT, etc.) as well as recognition in various applications. Since then several group members have progressed with their scientific careers and obtained the degree of doctor of technical sciences.

The natural need to exchange research experiences, inspiring scientific discussions not only on the domestic but also in the international field, has resulted in the idea of international scientific conferences under the name of Image Processing & Communications (IP&C).

IP&C dates back to 2009 when Institute of Telecommunications, UTP organized the first conference. It was one of the first initiatives to organize an Image Processing/computer Vision, Computer Science and Information and Communication Technology based conference in Poland with participation from multiple nations and universities. The proceeding of the first IP&C conference

© Springer Nature Switzerland AG 2019
M. Choraś and R. S. Choraś (Eds.): IP&C 2018, AISC 892, pp. 3–10, 2019.
https://doi.org/10.1007/978-3-030-03658-4_1

were published in the form of a book published by EXIT Akademicka Oficyna Wydawnicza. Then, in the next years thanks to the support of prof. Janusz Kacprzak - the editor of the series Advances in Intelligent Systems and Computing, the proceeding were published by Springer - as a part of this series. For IP&C conferences, always the rigorous double blind-review process was implemented that led to high-level articles that were accepted for presentation and publication in the book Image Processing & Communications Challenges. A selected number of manuscripts after further enhancement and review process were also included in one of journal special issues.

The paper is organized as follows: Sect. 2 contains reflection on the image processing aspects divided into five specific domains of application: biometrics, pattern recognition, applications of image processing, video processing and quality cases. The next section presents various aspects in communications such as cyber security and cloud computing. Conclusions and some plans for the future are provided in afterwards.

2   Image Processing

A picture is worth a thousand words. (Chinese proverb) We can distinguish four types of domains of application for the digital image processing (Fig. 1):

- restoration and enhancement image,
- image analysis,
- image coding with data compression,
- image synthesis.

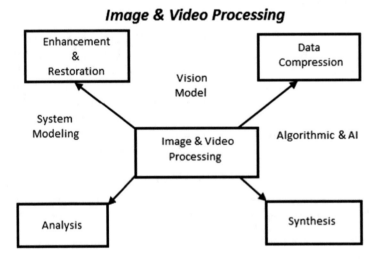

Fig. 1. Image & video processing

Image processing can be characterized not simply in terms of its domains of application, but also according to the nature of the results. There can be two types of input: an image or a description (Fig. 1).

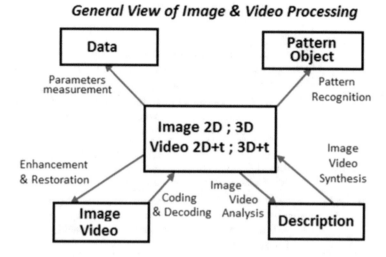

Fig. 2. General view of image & video processing

2.1 Applications

Image processing may be implemented in various applications discussed at IP&C, such as:

- the support of disabled persons - in [26] the interface for the head movement-based keyboard was proposed;
- the doctors support - features extraction improvement introduced in [33] for prostate cancer detection using images from the perfusion computed tomography;
- automatic waste sorting - based of a histogram of a waste photo, the decision PET/not-PET is given [4];
- understanding of natural images - in [32] a system for recognition the hand drawn flowcharts was proposed, that may support computer science engineers or mathematicians;
- entertainment - using GDL classifier in order to recognize the Karate movements which are performed faster than common-life gestures [12]

2.2 Biometrics

The key point of biometrics is to analyze the body parts images or a person's behavior in order to recognize or verify the identity. It was proposed more than 50 years ago but it has become an emerging challenge for researchers recently. The popularity of biometric solutions is increasing due to its wide applications (e.g. border control, biometric passports, forensic purposes, mobile phone unlocking, multi-factor authentication systems). Several biometric issues were present during each IP&C conference. Table 1 contains a set of the selected articles concerning about the biometrics (Fig. 2).

Table 1. Biometric research works

Feature	Description	Papers
Ear	Extraction moments form ear image, using 2D Fourier transform for features extraction and Euclidean distance for matching	[11]
Face	Using Wavelet transform and stochastic process for face-based recognition. Proposing a new method for facial features from colorful and frontal images	[3,30]
Fingerprint	Running Gabor-based features extraction on GPU	[21]
Iris and retina	Multimodal biometrics - 2D Gabor filters used to extract features from retina and iris	[8]
Keyboard typing	Using two times for identification - flight time (the time between releasing a key and pressing the second) and dwell time (the time when a specific key is in pressed state)	[29]
Knuckle	Applying PHT transform and SURF algorithm for features extraction	[5]
Mouth region	Calculating Radon features for lip-based biometrics and color moments and Gabor features for tongue-based approach	[6,7]
Palmprint	The comparison of pre-processing methods (blurs etc.) used in HOG-based verification system	[35]
Vascular	Using NIR-image and applying Discrete Fourier Transform	[15]

2.3 Pattern Recognition

Pattern recognition may be understood as a process of identification of the data regularities. Mostly it may be divided into some steps: data preprocessing, features extraction and classification (matching). Each of them has been improved by researchers for many years. In [10] authors focused on image preprocessing, image enhancement in details. The presented work concerns images of tree-rings that are used in dendrochronological measurements. In [17] the other preprocessing method is proposed - denoising. It was dedicated for reducing noise in computed tomography images. Features extraction was developed in [23], where Radon Transformation was used to find rotation and scale invariant features in order to support automatic recognition of post mail codes. In [20] lines detection was investigated. The proposed method was dedicated especially for robotics. In [31] a novel method of matching, which is based on the nearest neighbor searching, was proposed. Author in [9] presented the improvement of wide-implemented AdaBoost classifier.

2.4   Quality

Image quality assessment is considered as the crucial element in graphics and for image processing algorithms. In [25] the serious problem of image compression was mentioned. Even though cameras provide high quality images nowadays, users often save them as JPEG images and therefore they loose some quality. However, to analyze the quality some methods of assessment have to be provided: in [34] 2D images assessment, in [27] 3D images assessment and in [28] video frames assessment.

2.5   Video Processing

Due to the computer vision development, the video analysis has become increasingly important. However, the multimedia transmission requires wide bandwidth and storage space. Thus, some research concerning video coding and compression were presented [16, 18]. Apart of coding, the reliable object tracking seems to be the most yelling problem in terms of video analysis. It may be used in software for UAVs, autonomous cars, surveillance cameras. The possible approach was presented in [36].

3   Communications

Communication leads to community, that is, to understanding, intimacy and mutual valuing. (Rollo May)

From the beginning of IP&C various aspects of telecommunications and networks were discussed. Some of them concerned algorithms for optimization, routing, wide networks design, topologies as well as the practical aspects of networks realization etc. Recently, one of the emerging topic was cyber security and networks reslilience. The amount of the available information, processed and exchanged via Internet has increased rapidly recently. Due to this fact, the security of the data online become a key point for providers and researchers. Authors in [13] analyzed the thread of digital sabotage to the critical infrastructure (e.g. power plants, airports). Intrusion detection and prevention systems are currently one of the most important mechanisms implemented in order to improve security in the network:

- in [2] the DDoS attack was detected using a novel greedy 1D KSVD algorithm,
- in [1] there was presented a statistical model called ARFIMA (Autoregressive Fractional Integrated Moving Average) for anomaly detection in the network traffic,
- in [19] Extreme Machine Learning for anomaly detection was adopted.

Another aspect of modern networks is cloud, edge and cluster computing. Cloud computing was introduced at the beginning of XXI century. However, the popularity of cloud and fog solutions has exploded recently. Authors in [14] proposed a simulator for cloud computing infrastructure. The simulator is considered to be useful in cloud computing research (like cloud efficiency testing).

In [24] cloud computing is proposed as a scenario for virtualization system (services like IaaS, PaaS, SaaS) dedicated for students laboratories at universities. One of the biggest challenges in cloud computing is isolating the consumer data and accessing it. In [22] this problem was solved by implementing the TLS/SNI solution.

4   Conclusions

In this paper we reflected on the anniversary of Image Processing and Communications Conference. It is now a well known venue for discussing current scientific and practical aspects in a nice and quite loose atmosphere. The conference is also well know for being friendly and supportive to young researchers, and also for interesting social events. We hope to organize it in the next years as well, since the number of relevant interesting research topics still grows.

References

1. Andrysiak, T., Saganowski, L.: Network anomaly detection based on ARFIMA model. In: Choraś, R.S. (ed.) Image Processing and Communications Challenges 6. AISC, vol. 313, pp. 255–262. Springer, Cham (2015)
2. Andrysiak, T., Saganowski, L., Choraś, M.: DDoS attacks detection by means of greedy algorithms. In: Choraś, R.S. (ed.) Image Processing and Communications Challenges 4. AISC, vol. 184, pp. 303–310. Springer, Cham (2013)
3. Bobulski, J.: 2DHMM-based face recognition method. In: Choraś, R.S. (ed.) Image Processing and Communications Challenges 7. AISC, vol. 389, pp. 11–18. Springer, Cham (2016)
4. Bobulski, J., Piątkowski, J.: PET waste classification method and plastic waste DataBase - WaDaBa. In: Choraś, M., Choraś, R.S. (eds.) Image Processing and Communications Challenges 9, pp. 43–48. Springer, Cham (2018)
5. Choraś, M., Kozik, R.: Knuckle Biometrics for Human Identification. In: Choraś, R.S. (ed.) Image Processing and Communications Challenges 2. AISC, vol. 84, pp. 91–98. Springer, Cham (2010)
6. Choraś, R.S.: Automatic tongue recognition based on color and textural features. In: Choraś, R.S. (ed.) Image Processing and Communications Challenges 8. AISC, vol. 525, pp. 16–26. Springer, Cham (2017)
7. Choraś, R.S.: Lip-prints feature extraction and recognition. In: Choraś, R.S. (ed.) Image Processing and Communications Challenges 3. AISC, vol. 102, pp. 33–42. Springer, Cham (2011)
8. Choraś, R.S.: Multimodal biometric personal authentication integrating iris and retina images. In: Choraś, R.S. (ed.) Image Processing and Communications Challenges 2. AISC, vol. 84, pp. 121–131. Springer, Cham (2010)
9. Dembski, J.: Multiclass AdaBoost classifier parameter adaptation for pattern recognition. In: Choraś, R.S. (ed.) Image Processing and Communications Challenges 8. AISC, vol. 525, pp. 203–212. Springer, Cham (2017)
10. Fabijańska, A., Danek, M., Barniak, J., Piórkowski, A.: A comparative study of image enhancement methods in tree-ring analysis. In: Choraś, R.S. (ed.) Image Processing and Communications Challenges 8. AISC, vol. 525, pp. 69–78. Springer, Cham (2017)

11. Frejlichowski D.: Application of the Polar-Fourier greyscale descriptor to the problem of identification of persons based on ear images. In: Choraś, R.S. (eds.) Image Processing and Communications Challenges 3. AISC, vol. 102, pp. 5–12. Springer, Cham (2011)
12. Hachaj, T., Ogiela, M.R., Piekarczyk, M.: Real-time recognition of selected karate techniques using GDL approach. In: Choraś, R.S. (ed.) Image Processing and Communications Challenges 5. AISC, vol. 233, pp. 98–106. Springer, Cham (2014)
13. Hald, S.L.N., Pedersen, J.M.: The threat of digital hacker sabotage to critical infrastructures. In: Choraś, R.S. (ed.) Image Processing and Communications Challenges 5. AISC, vol. 233, pp. 98–106. Springer, Heidelberg (2014)
14. Hanczewski, S., Kędzierska, M., Piechowiak, M.: A simulator concept for cloud computing infrastructure. In: Choraś, R.S. (ed.) Image Processing and Communications Challenges 6. AISC, vol. 313, pp. 257–264. Springer, Cham (2015)
15. Kabaciński, R., Kowalski, M.: Human vein pattern segmentation from low quality images - a comparison of methods. In: Choraś, R.S. (ed.) Image Processing and Communications Challenges 2. AISC, vol. 84, pp. 105–112. Springer, Heidelberg (2010)
16. Karwowski, D.: Improved adaptive arithmetic coding in MPEG-4 AVC/H.264 video compression standard. In: Choraś, R.S. (ed.) Image Processing and Communications Challenges 3. AISC, vol. 102, pp. 257–264. Springer, Heidelberg (2011)
17. Knas, M., Cierniak, R.: Computed tomography images denoising with Markov Random Field Model parametrized by prewitt mask. In: Choraś, R.S. (ed.) Image Processing and Communications Challenges 6. AISC, vol. 313, pp. 53–58. Springer, Cham (2015)
18. Knop, M., Dobosz, P.: Neural video compression algorithm. In: Choraś, R.S. (ed.) Image Processing and Communications Challenges 6, AISC, vol. 313, pp. 257–264. Springer, Cham (2015)
19. Kozik, R., Choraś, M., Holubowicz, W., Renk, R.: Extreme learning machines for web layer anomaly detection. In: Choraś, R.S. (ed.) Image Processing and Communications Challenges 8, AISC, vol. 525, pp. 271–278. Springer, Cham (2017)
20. Lech, P., Okarma, K., Fastowicz, J.: Fast machine vision line detection for mobile robot navigation in dark environments. In: Choraś, R.S. (ed.) Image Processing and Communications Challenges 7. AISC, vol. 389, pp. 151–158. Springer, Cham (2016)
21. Lehtihet, R., El Oraiby, W., Benmohammed, M.: Improved fingerprint enhancement performance via GPU programming. In: Choraś, R.S. (ed.) Image Processing and Communications Challenges 3, AISC, vol. 102, pp. 13–22. Springer, Heidelberg (2011)
22. Łaskawiec, S., Choraś, M.: Considering service name indication for multi-tenancy routing in cloud environments. In: Choraś, R.S. (ed.) Image Processing and Communications Challenges 8. AISC, vol. 525, pp. 271–278. Springer, Cham (2017)
23. Marchewka, A., Miciak, M.: The feature extraction from the parameter space. In: Choraś, R.S. (ed.) Image Processing and Communications Challenges 8. AISC, vol. 525, pp. 144–153. Springer, Cham (2017)
24. Munoz-Exposito, J.E., de Prado R.P., Garcia-Galan, S., Rodriguez-Reche, R., Marchewka, A.: Analysis and real implementation of a cloud infrastructure for computing laboratories virtualization. In: Choraś, R.S. (ed.) Image Processing and Communications Challenges 7. AISC, vol. 389, pp. 275–280. Springer, Cham (2016)
25. Nowsielski, A.: Quality prediction of compressed images via classification. In: Choraś, R.S. (ed.) Image Processing and Communications Challenges 8, AISC, vol. 525, pp. 35–42. Springer, Cham (2017)

26. Nowsielski, A.: Swipe-like text entry by head movements and a single row keyboard. In: Choraś, R.S. (ed.) Image Processing and Communications Challenges 8. AISC, vol. 525, pp. 136–143. Springer, Cham (2017)
27. Okarma, K.: On the usefulness of combined metrics for 3D image quality assessment. In: Choraś, R.S. (ed.) Image Processing and Communications Challenges 6. AISC, vol. 313, pp. 137–144. Springer, Cham (2015)
28. Okarma, K.: Video quality assessment using the combined full-reference approach. In: Choraś, R.S. (ed.) Image Processing and Communications Challenges 2. AISC, vol. 84, pp. 51–58. Springer, Heidelberg (2010)
29. Panasiuk, P., Saeed, K.: A modified algorithm for user identification by his typing on the keyboard. In: Choraś, R.S. (ed.) Image Processing and Communications Challenges 2. AISC, vol. 84, pp. 121–131. Springer, Heidelberg (2010)
30. Papaj, M., Czyżewski, A.: Facial Features Extraction for Color, Frontal Images. In: Choraś, R.S. (ed.) Image Processing and Communications Challenges 3, AISC, vol. 102, pp. 23–32. Springer, Heidelberg (2011)
31. Swiercz, M., Iwanowski, M., Sarwas, G., Cacko, A.: Combining multiple nearest-neighbor searches for multiscale feature point matching. In: Choraś, R.S. (ed.) Image Processing and Communications Challenges 7. AISC, vol. 389, pp. 231–238. Springer, Cham (2016)
32. Szwoch, W., Mucha, M.: Recognition of hand drawn flowcharts. In: Choraś, R.S. (ed.) Image Processing and Communications Challenges 4. AISC, vol. 184, pp. 65–72. Springer, Heidelberg (2013)
33. Śmietański, J.: The usefulness of textural features in prostate cancer diagnosis. In: Choraś, R.S. (ed.) Image Processing and Communications Challenges 2. AISC, vol. 84, pp. 223–229. Springer, Heidelberg (2010)
34. Tichonov, J., Kurasova, O., Filatovas, E.: Quality prediction of compressed images via classification. In: Choraś, R.S. (ed.) Image Processing and Communications Challenges 8. AISC, vol. 525, pp. 136–143. Springer, Cham (2017)
35. Wojciechowska, A., Choraś, M., Kozik, R.: Evaluation of the pre-processing methods in image-based palmprint biometrics. In: Choraś, M., Choraś, R.S. (eds.) Image Processing and Communications Challenges 9, pp. 43–48. Springer, Cham (2018)
36. Zawistowski, J., Garbat, P., Ziubiński, P.: Multi-object tracking system. In: Choraś, R.S. (ed.) Image Processing and Communications Challenges 5. AISC, vol. 233, pp. 98–106. Springer, Heidelberg (2014)

Development of a Mobile Robot Prototype Based on an Embedded System for Mapping Generation and Path Planning - Image Processing & Communication - IPC 2018

Enrique Garcia[1]([⊠]), Joel Contreras[1]([⊠]), Edson Olmedo[1]([⊠]),
Hector Vargas[1]([⊠]), Luis Rosales[2]([⊠]), and Filiberto Candia[3]([⊠])

[1] Faculty of Electronics, Universidad Popular Autonoma del Estado de Puebla,
21 Sur 1103, Barrio de Santiago, 72410 Puebla, Mexico
enriquerafael.garcia@upaep.edu.mx,
{joel.contreras,edson.olmedo,hectorsimon.vargas}@upaep.mx
[2] Faculty of Postgraduate Studies, Universidad Popular Autonoma del Estado
de Puebla, 21 Sur 1103, Barrio de Santiago, 72410 Puebla, Mexico
luis.rosales@upaep.mx
[3] Faculty of Engineering, Benemerita Universidad Autonoma de Puebla,
Boulevard Valsequillo, San Manuel, 72570 Puebla, Mexico
filinc@hotmail.com

Abstract. The project approach in teaching is now considered a promising method that relates the learning process to the conditions of solving real problems. Student interest stimulates in the topic under study, giving them differentiated skills, including those that are related to teamwork. This also allows students with different interests and levels of preparation to work simultaneously on the solution of one problem. Application of the project approach for training in complex new areas requires trained equipment and specialists. The specialists direct the process of creative search for students in gaining relevant knowledge, as well as specifics in determining the problem under study.

Keywords: Mobile robot · 2D SLAM · Path planning · Prototype Embedded system

1 Introduction

This prototype is a mobile robot that allows experimenting with different integrated elements, study the work of a robot and utilize algorithms to master the robotic areas. The advantages of the proposed approach are:

- Interdisciplinary and direct interconnection between elements within a single project

© Springer Nature Switzerland AG 2019
M. Choraś and R. S. Choraś (Eds.): IP&C 2018, AISC 892, pp. 11–19, 2019.
https://doi.org/10.1007/978-3-030-03658-4_2

– Involving students in solving actual scientific and technical problems
– Familiarize with some of the contemporary problems of their region, together
  with the possibility of studying standard academic disciplines in the framework of the project approach.

Nowadays, the need for technological advancement in universities [1] of developing countries is greater because technology enables students to become much
more involved in the construction of their own knowledge. Cognitive studies
show this capacity is key for successful learning. Also the intensive development
of robotics allows educational institutions to open departments and faculties
for training and research in the field of technical cybernetics and robotics. Full-
fledged training of students about design and programming of robots requires an
active model. This model will allow studying the design features of small-wheeled
vehicles, the construction of microprocessor control systems, and to investigate
the operation of computer vision systems. The object of study can be the analysis of data from the map environment, inertial navigation systems, or navigation
chambers.

The computer system of the robot allows updating of firmware [2]; this property makes it possible to carry out research work on the creation of software
for cybernetic motion control and autonomous movement systems. The result of
such research is scientific work in the fields of robotics, technical cybernetics or
applied mathematics.

2   Conceptual Design

The first approach provides a prototype for academic and research purposes on a
cost-effective robot that can operate both indoors and outdoors, equipped with
sensors which let it work to cover mapping activities and implement computer
vision algorithms. Even though kits exist for building open-source robots there
is still a need to improve and manage modern technology to make robotics a
more efficient field in avoiding big losses while performing a task considering the
lowest price. Underdeveloped countries such as Mexico certainly need to find
solutions to avoid the risk of losing economic resources, which is why it is very
unlikely to acquire marketable robots if the venture is high. Computer aided
design software gives the thrust to test the robot's functionality without having
to do it physically via FEA (Finite element analysis). The study eventually
identifies how to integrate from a weight budget set in the beginning (Fig. 1).

3   Hardware and Software Design

Given the fact of being a low-cost robot, the usage of cheap yet effective components and circuits is one of the main characteristics of the design. The idea of
integrating parts available on the market is crucial in the case that spare parts
will be required. Cost and quality is another pillar because there is no advantage
in deficient performance even if the cost is acceptable. The hardware design is
split into five parts:

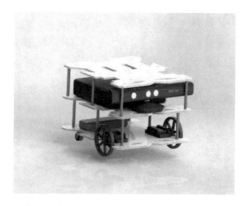

Fig. 1. CAD model with no arm on it

3.1 Mechanics

A symmetric design was chosen as the first model to test versatility and flexibility to adapt to another model if needed with almost no remaining impact on the budget. The idea of developing a four-wheel robot with traction control only in two of the wheels was selected as the most effective way to save resources. As a result, two electric servo motors were assembled next to the main wheels to provide movement. Motors are mounted at the bottom of a base made of acrylic and coupled with 3D printed parts using ABS (Acrylonitrile Butadiene Styrene) material. The robot is formed by four acrylic bases. The lowest layer is the base link and connects traction with the robot. The first space between the first and second layer is designed to manage the electronics. The second space between the second and third layer is intended for Kinect v1 sensors and finally the upper level is aimed at the actuators. A limitation found on the current prototype is that the wheels lacking traction did not ensure an efficient movement and thus swivel wheels were integrated which improve stability.

3.2 Electronics

A Raspberry Pi 3 and a 14.4 V LiPo (Lithium polymer) battery and a power bank with designed harnesses is equipped in the robot. The idea to separate the power system from the electronics is to guarantee the success of any mission and avoiding the risk of damage in low level circuits exposed to power levels if they somehow fail. The power bank was selected to power the electronics while the LiPo battery was provided to control the power system. A 12 V regulator is used to handle the Kinect sensor, no regulation is used for the motors, and a 5 V regulator handles a robotic arm equipped with four small servo motors. The robotic arm is controlled through PWM (Pulse Width Modulation) signals as well as the motors for the robot's wheels. The 3D processing is performed by the Kinect sensor itself (Fig. 2).

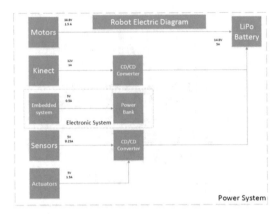

Fig. 2. General schematic

3.3 Communications

The Raspberry Pi 3 module works as the onboard computer. Data is sent via a Wi-Fi network adapter integrated in the embedded system. Afterwards, the information is received in a server the computer retrieves to save and post-process the raw data. Port-forwarding technique was implemented given the number of advantages of it. As it has been documented it is possible to connect an HDMI-based (High-Definition Multimedia Interface) system directly to the microcontroller, but is not recommended. String commands are sent from onboard computer to server through a TCP/IP protocol, where in fact the most common protocol used is TCPROS based on standard TCP/IP sockets.

Master PC: A workstation computer running Ubuntu has ROS installed on it. Given the low capacity of the onboard computer, information is sent to the master PC or server and then post analyzed. If there is only LAN connection then the system will work but without remote control. If there is a WAN connection then by using a port forwarding technique it is possible to let workflow from the outside of the network.

The robot has two main methods of action: the first being an autonomous system and the second being remotely controlled. The idea behind this is firstly for an operator to control it and verify the field that will be subject to analysis. In case of losing connection, then the robot will work autonomously to carry out the task if possible.

3.4 Transducers

Sensors: The robot is equipped with a Kinect sensor, a HD (High quality) endoscope camera, and two HC-SR04 ultrasonic sensors attached to an ADC (analog to digital) converter connected to Raspberry GPIO (General Purpose Input/Output) pins. The results have shown a detection range from 430 mm to almost 6000 mm given by the Kinect sensor.

Actuator: There is a robotic arm working as the main robot's actuator consisting of five PWM controlled servo motors.

The HD endoscope camera was mounted to help robotic arm manipulation in the event of remote control. According to each mission or task, it is possible to mount or unmount different kinds of sensors and actuators.

3.5 Software

The onboard computer (Raspberry Pi 3) has been uploaded with an Ubuntu system that will provide the integration supported by ROS (Robotic Operating System) which will serve for the core functions. Both are upgraded to the latest versions. ROS structure mainly covers topics such as:

- Packages: Main units that may contain a process or node. These are useful for library settings, configuration files, datasets and so on.
- Nodes: It is defined as a process that performs a computation. They are written in C++ and are the ones that control each activity that the robot will perform such as 3D laser scanning, mapping, manipulation, computer vision, odometry, integration, etc.
- Master: It provides an interconnected communication among all nodes. At least one master needs to be running to let the system be active.
- Bags: Files containing ROS information. They are useful for saving and playing ROS data. Storing sensor data is done by this mechanism.

4 Advanced Algorithms

4.1 Motion

Odometry algorithms are implemented to ensure motion and information given by encoders is reported every 500 ms. A self-evaluation frame provides the displacement and angle of the robot for indicating change of path or movement for a differential-drive mobile robot. All estimation was calculated according to the kinematic model, given by:

$$v_{final} = \frac{v_{left} + v_{right}}{2} = \frac{\omega_{left}R + \omega_{right}R}{2}$$

where:

v_{final}: is the resulting final velocity
v_{left}: is the left velocity
v_{right}: is the right velocity
$\omega_{left}R$: is the left angular velocity
$\omega_{right}R$: is the right angular velocity

Some constraints are related to the mechanical implementation of the robot, while some systematic errors are specific and remain constant, and a common error is due to the difference in size of the wheels which can be summarized as the diameter inequality [3]

$$E = \frac{D_1}{D_2}$$

where:

E: is the obtained error
D_1: is the diameter of the first wheel
D_2: is the diameter of the second wheel

Apart from straight line motion, there is an error when turning which in this project has been mitigated through the implementation of an IMU (Inertial measurement unit) sensor.

4.2   Mapping

This system uses SLAM (Simultaneous Localization and Mapping) as a way of solving the mapping problem [4]. No GPS sensor is equipped but it will soon be part of the main system to improve efficiency. To this point the robot can navigate and build a map of an unknown place. Two different algorithms were chosen and improved by modifying initial conditions. Gmapping (for testing) and hector_slam packages from ROS are the based mapping algorithms. A feedback mechanism has been proposed however it is still in progress. The Kinect sensor works as a pseudo laser to create a 2D map of the environment, but also works as a 3D map thanks to point cloud data. However, it does not run the entirety of the time because it requires excessive storage space.

4.3   Path Planning

For these purposes a 2D space was created. In the first attempt, with no IMU sensor attached, a single XY plane is represented and limited by the Ackermann configuration [5]. As stated before, a map is a mandatory requirement in order to establish the best route. This chosen algorithm for path planning is the Dijkstra algorithm [6] because of its advantages such as good performance and simplicity. The method is simple: it first sets an amount of nodes where the robot is going to navigate and then it gives a value to each one of them.

4.4   Artificial Vision

OpenCV is a software for computer vision and can be integrated to a ROS system. So far, basic line detection and object recognition can work satisfactory, however advanced algorithms have been tested to improve results for manipulation, mapping and motion. A point cloud has been generated to a store in an external device for post-processing, although there is still a lot of research needed for further improvements.

5 Implementation and Integration

One of the most problematic stages for robotics is when integration and implementation must be met. This project has shown good results for controlling, mapping, path planning and computer vision wirelessly. ROS has demonstrated to be a middleware to communicate a master and slave system easily while at the same time sending commands and receiving data remotely. This robot is now conceived as an academic platform for research. In the future, it will be part of a beginning of new missions for different kinds of applications such as exploration of unknown or dangerous environments and academic topics (Fig. 3).

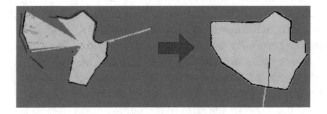

Fig. 3. Generated map

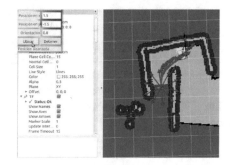

Fig. 4. Path planning implementation

This prototype has been used as a medium for generating maps for the different classrooms which belong to the laboratory of control of the university. Images described before show both the map generation and the path planning implementation, the results have been compared with manual measurements giving good results so far. The device is now used as a tool of learning for students interested in robotics and may be improved to give other skills to whomever needs it (Fig. 4).

6 Conclusions and Future Work

Both map generation and path planning was implanted successfully by using an embedded system. After some tests on the same location it was possible to identify small variation, the user can control the robot to generate a map and after that point then the robot is able to reach one or some points according to the needs. Time is not a major issue and even dynamic obstacles can be understood by the robot in order to create an efficient route. This project is part of bigger plan to develop a full service robot designed for home tasks. C++ programming integrated on ROS has also showed an optimum result which can be easy to adjust and improve. So far this robot is able to perform 2D SLAM by using an embedded system however there is still an issue related to 3D SLAM, given the fact that there is a Kinect sensor mounted on the platform then a 3D map is a major challenge. Autonomous map generation has not been implemented however there have been some trials to perform autonomous navigation so in the future there is a high chance to be integrated. Sensor integration for specific tasks is also an innovation that can be designed and integrated. A fundamental part of the results was for the students have a testbed where they can design experiments using engineering applications almost immediately because the mechanical structure and the sensors are commercially off the shelf (COTS) and do not require development. Thus, students require only an introduction to programming and ability to perform less complex tasks. These are the achievements that are observed in the development of the applications derived from the use of the mobile robot that has helped us to teach students. Both map generation and path planning were successfully encoded by using an embedded system. After testing in the same location it was possible for the mobile robots to identify small variations in their environment while the user also controlled the robot to generate a map. The next progression was for the robot to able to reach one or more specified points according to the needs. The time required to accomplish these tasks is not a major issue and even dynamic obstacles can be understood by the robot in order to create an efficient route with C++ programming integrated into ROS which has shown optimum results and can easily be adjust and improved.

Now we have different areas of opportunity to improve some specific tasks such as autonomous map generation which has not been implemented. There have, however, been trials to perform autonomous navigation so in the future there is a high chance of integration. Sensor integration for specific tasks is also an innovation that can be designed. Thus far, this robot can perform 2D SLAM by using an embedded system however there is still an issue related to 3D SLAM, given the fact that there is a Kinect sensor mounted on the platform making 3D map a major challenge.

We must take into account that this project is part of bigger plan to develop a full-service robot designed for assisting with home-based tasks and that the advantage of creating a development platform to carry out our tests requirements helps us identify possible improvements and/or correct the ones we have already identified.

Acknowledgments. This study was mainly supported by an internal budget from Universidad Popular Autonoma del Estado de Puebla (UPAEP), a non-profit university located in Puebla, Mexico in the year 2018 as a part of a continuing education project for students from the Electronics department. We also appreciate the support from the National Council of Science and Technology of Mexico (CONACYT) which has granted scholarships for postgraduate studies.

References

1. Gonzalves, J., Lima, J., Costa, P., Moreira, A.: Manufacturing education and training resorting to a new mobile robot competition. In: Flexible Automation Intelligent Manufacturing 2012 (Faim 12) Ferry Cruise Conference Helsinki-Stockholm-Helsinki (2012)
2. Tatur, M.M., Dadykin, A.K., Kurdi, M.M.: Multifunction system of mobile robotics. In: 2016 Third International Conference on Electrical, Electronics, Computer Engineering and Their Applications (EECEA), pp. 110–113. IEEE (2016)
3. Krinkin, K., Stotskaya, E., Stotskiy, Y.: Design and implementation Raspberry Pi-based omni-wheel mobile robot. In: Artificial Intelligence and Natural Language and Information Extraction, Social Media and Web Search FRUCT Conference (AINL-ISMW FRUCT), pp. 39–45. IEEE, November 2015
4. Grisetti, G., Stachniss, C., Burgard, W.: Improved techniques for grid mapping with Rao-Blackwellized particle filters. IEEE Trans. Robot. **23**(1), 34–46 (2007)
5. Khristamto, M., Praptijanto, A., Kaleg, S.: Measuring geometric and kinematic properties to design steering axis to angle turn of the electric golf car. Energy Procedia **68**, 463–470 (2015)
6. Ben-Ari, M., Mondada, F.: Mapping-based navigation. In: Elements of Robotics, pp. 165–178. Springer, Cham (2018)

Using Stereo-photogrammetry for Interiors Reconstruction in 3D Game Development

Mariusz Szwoch[1]([✉]), Adam L. Kaczmarek[1], and Dariusz Bartoszewski[2]

[1] Gdansk University of Technology, Gdansk, Poland
szwoch@eti.pg.edu.pl, adam.l.kaczmarek@eti.pg.edu.pl
[2] Forever Entertainment s.a., Gdynia, Poland
dariusz.bartoszewski@forever-entertainment.com

Abstract. This paper proposes a new approach to the reconstruction of building interiors based on stereo-photogrammetry. The proposed technology can be used, among others, for modeling rooms for video games whose action takes place in locations reflecting real interiors. The original, large-scale experiment showed that the proposed approach is economically justified in the case of larger and more complex spaces which are more difficult to model in a traditional way. Also novel approach is proposed to deal with the problem of scanning flat uniform surfaces.

1 Introduction

Modern 3D video games require hard use of high quality 3D assets for both game characters as well as level scenery. In the traditional approach, all such assets are created by 3D artists or bought from a store. The first approach is a very labor-intensive task, while the second one does not guarantee originality of the assets, what is easily recognized by players. Situation becomes more complicated in case when game concept requires that the action takes place in a location that represents a real world place. In such case, traditional, handmade modeling seems to be non-creative, yet time and labor consuming task that could be replaced by automatic or at least semi-automatic machine work. Automation of room reconstruction not only reduces time and the cost of modeling the 3D stage, but also supports rapid prototyping, which is a very important element of agile development process and often determines the further direction of the game development. That is why, it is so important to accelerate the process of modeling 3D scenes that have the appearance as close as possible to the target. As long as this target appearance is a figment of the artist's imagination, the only solution is to improve the 3D modeling tools and hire skilled 3D artist. If, however, objects appear on the scene to reflect real objects, then there is the possibility of using additional tools allowing for automatic or semi-automatic reconstruction of these objects for the needs of the game.

M. Choraś and R. S. Choraś (Eds.): IP&C 2018, AISC 892, pp. 20–29, 2019.
https://doi.org/10.1007/978-3-030-03658-4_3

There are several possible solutions of reconstruction of 3D inanimate objects, including photogrammetry [2, 7], methods based on lasers and augmented depth sensors, e.g. structured light RGB-D ones [5,10,13]. Unfortunately, all of the mentioned solutions have some drawbacks that limit their usage in video game industry, especially by small and medium developer teams. Laser based scanners are very precise as to geometry scanning but are very expensive and do not support textured models. Time-of-Flight (ToF) sensors are cheaper but have very limited resolution and still do not provide texture information. RGB-D scanners, based mostly on the structured infrared light patterns, like Microsoft Kinect or Intel RealSense, can provide complete 3D models of usually enough quality but are limited to small distances of about several meters. Finally, photogrammetry methods can provide high quality textured models but requires numerous input data and high computational costs. Moreover, it often fails to reconstruct flat surfaces with no salient points.

In this paper, the main concept of the STERIO project is presented, which aims to overcome problems of above approaches by using stereo-photogrammetry methods which, under some assumptions, can offer reliable scanning of textured objects at reasonable financial and computational costs. Performed experiments indicate that such approach is faster than traditional one. Additionally, a special way of treating flat surfaces is proposed, which uses graphic patterns displayed on a problematic surface.

2   Background

Three-dimensional (3D) video games are the most important and the most spectacular part of the video games market. They dominate primarily on the platforms of personal computers and video consoles, but they are also an essential and still growing part of games for mobile and browser platforms.

Because independent modeling of three-dimensional objects is very time-consuming, for some time the methods, algorithms, tools and equipment supporting the 3D scanning have been developed. Initially, they focused on scanning small and medium 3D objects, including motion capture systems. To scan such smaller objects, relatively cheap camera sets and various types of 3D scanners and depth sensors could be used, like Kinect or RealSense. Modeling of larger objects and 3D locations could be carried out mainly using expensive laser scanners which are available mainly to larger development studios. With a smaller game budget, the modeling of larger assets remained the domain of 3D graphic designers. It is only recently that methods allowing the use of cameras for scanning buildings, their interiors and even vegetation have attracted a lot of interest [3, 5–7, 9].

3D scanning technology has been developed for the last decades and used in different fields, such as preserving objects significant for a cultural heritage [3]. The best solution to acquire relatively small objects are 3D scanners using different scanning technologies, e.g. 3D structural-light scanners [6]. Unfortunately, due to their limitations, such 3D scanners cannot be at reasonably costs used for scanning bigger objects, e.g. buildings and their interiors.

This problem is less significant when scanning devices based on lasers are used. This class of devices includes *time-of-flight* (TOF) cameras and devices based on *light detection and ranging* (LIDAR) technologies [9,11]. LIDAR has a movable laser which is pointed in different directions during scanning. A 3D scan results from a series of measurements performed around a single point. TOF makes a scan with the use of light laser beam distracted in different directions. The disadvantage of both of these devices is such that the resulting 3D scan is sparse and lacks color information. LIDARs and TOF cameras are also relatively expensive. However, these scanners are more immune to light conditions than 3D structural-light scanners.

3D scans can be also obtained with the use of standard, optical cameras which are very accessible and relatively cheap. In general, there are two kinds of methods for acquiring 3D scans with cameras, which are *photogrammetry* and *stereo-photogrammetry*. The photogrammetric approach bases on a series of usually unrelated photos taken from around a scanned object. Special algorithms, such as *scale-invariant feature transform* (SIFT) and *speeded up robust features* (SURF) are used to detect the characteristic points, or *landmarks*, in those images [4]. Matching algorithms, such as *random sample consensus* (RANSAC) are then used to find the correspondence between images pair. On this base, algorithms, such as *structure from motion* (SfM), allow to reconstruct a 3D point cloud representing the scanned object. This approach is the most popular nowadays as it allows for the reconstruction of buildings, plants, and other objects based on uncorrelated images possible taken using different cameras at different time, e.g. as is the case in social media services. This approach is also used in a professional tools, such as Agisoft Photoscan [1]. The greatest disadvantage of this approach is the computational power needed to process big sets of images.

Another technique, which allows to retrieve a 3D view of objects, is stereo-photogrammetry that belongs to the same class of technologies as stereo vision [7,8]. In this method, a rigid set of two cameras is used, called a *stereo-camera*. As the relative position of cameras is fixed, it is possible to reconstruct partial point cloud of the scanned object directly from a stereo-image, simplifying the calculations time. This can be an essential advantage over SfM in some application area such as rapid modeling and 3D scenes prototyping in video games. Although, this approach requires using a tripod, it can become another advantage in some situations, such as insufficient illumination or hard-to-process surfaces, when it is possible to take several photos from exactly the same location in different lighting conditions [12].

3   The STERIO Project

The goal of the STERIO project is to develop an efficient technology and programming tools for the effective development of 3D models using stereoscopic photography [12]. The results of the project will allow for faithful recreation of real world scenery in the virtual world of video games. Applying the proposed technology will also allow for more effective and less costly location generation for games which action takes place in real locations. Such approach will have a big advantage over the indirectly competitive 3D laser-scanning technology as it produces a similar effect with fewer resources and shorter production time. To facilitate operation, the technology will first be integrated with the most popular Unity 3D engine.

3.1   The STERIO Data Set

In the first stage of the STERIO project a comprehensive data set of images was prepared, allowing for further research in different aspects of 3D scanning of interiors for video games application. The whole STERIO dataset consists of over 10 thousand images of 25 various test locations and scenes that can be used for validation of developed algorithms. For each location, a different number of data subsets were prepared, consisting of images taken under different lighting conditions with different exposition parameters, camera location and orientation points, additional visual markers displayed, and other experiment setting. Some datasets are accompanied by additional images of a special chessboard pattern for cameras calibration. Sample images from the STERIO dataset are presented in Fig. 1.

Fig. 1. Sample images from the STERIO dataset: modern interior (an office) with additional visual pattern displayed, a corridor from an abandoned manor, a university hall, a restaurant's cellar

3.2   Reconstruction of Smooth Surfaces Using Displayed Patterns

Reconstruction of flat, uniform and monochromatic surfaces with no landmarks, or salient points, is one of the main problems of photogrammetry. Such surfaces, which are often found in interiors (e.g. walls and ceilings), are most often not reproduced by existing applications and libraries supporting photogrammetry (e.g. [1]), resulting in the discontinuity, or a hole, in the object's grid. There are several approaches possible to solve this problem. It is possible to place some markers on such surfaces, which is rather inconvenient, time-consuming and do not guarantee the final success, as there are still no landmarks between the markers. Another approach is to manually model such improperly reconstructed surfaces. In the STERIO project, we propose a novel approach to overcome this issue and increase the completeness of automatically obtained 3D model. In this approach, a special pattern is displayed on such planes (Fig. 1a), which provides details necessary for the proper scanning of the surface. This approach is not possible or at least not practical when using popular photogrammetric approach, as the displayed pattern overlaps the original texture of the surface. To avoid this undesirable effect, we propose to use the stereo-photogrammetry for such surfaces, in which two stereo-pairs are taken from the same stereo-kit position and orientation - one with the pattern displayed and one - without it. A set of images containing displayed pattern are used to reconstruct the interior's geometry while the other images are used to restore interior's original material, or texture.

Because stereo-photogrammetry with two cameras forces the use of a photographic tripod, in the applied approach there is no problem with maintaining the same position and orientation of both cameras and the fitting of photos from both sets, which is not possible when using individual photos in the photogrammetry process. There are two drawbacks of the presented approach. Firstly, it requires frequent switching on and off the LCD projector used to project a pattern, which can be easily omitted, e.g. by 'displaying' the black image. The second is the necessity of additional texture processing in order to restore their original colors. Fortunately, the proposed extension does not complicate or significantly extend the time of the whole process of interiors scanning. The initial results of performed experiments, presented in the next section, are very encouraging.

4   Experiments

Two experiments were carried out to validate the suitability of the use of stereo-photogrammetry in reconstruction of interiors for video game purposes. In the first experiment (Sect. 4.1), several different interiors were reconstructed by the traditional modeling and the proposed approach. This experiment is pioneering on a global scale and no such comparison has been published yet. In the next experiment (Sect. 4.2) several different patterns were tested to indicate which is the best one in this specific application.

4.1 Evaluation of the Profits from Using Photogrammetric Approach in Modeling Interiors

The goal of this experiment was to compare the time-complexity of modeling of the selected rooms using classic and photogrammetry-based approaches. The comparison made will allow estimating the possible benefits related to the development of the technology proposed within the project. As there are no such comparisons available, neither in the scientific nor industry literature, the results of this experiment are very important for further realization of the STERIO and similar projects.

In the case of the classic approach, models are created on the basis of a set of photos and basic distance measurements allowing for appropriate scaling of the model. In order to obtain the models based on photogrammetry, the Agisoft Photoscan software was used [1], which was considered as one of the best software available in the market at the moment. In both cases, the final stage of the work was to adapt the models to the requirements of the Unity environment and video games. In order to increase the reliability of the obtained results, the task was broken up into subtasks performed by different 3D graphic designers.

The experiment assumption was to model six varied interiors (Fig. 2) for a video game purposes using both approaches and compare their total time and estimated costs. In the first approach, three computer graphic designers were supposed to traditionally model two different interiors each, using a series of photos and very simplified models as a scale reference. All geometry was modeled by hand and textures were created or adopted from available assets. Only room specific textures, such as paintings or frescos, were prepared based on photos. Final models were optimized for using within game engines and finally imported into Unity as separate interactive scenes.

Fig. 2. Six interiors modeled within the experiment: Room3, Room4, Room1, Small-Room, Parking, and the Corridor (finally not used in the comparison)

In the second approach, each graphic designer was to model interiors, processed in the first stage by another one, using results of photogrammetry-based 3D scanning with Photoscan. As 3D scans received from Photoscan has numerous imperfections they cannot be directly used in video games. Graphic designers had to manually reduce complexity of the models mesh and textures, correct bad shapes, conceal defects and fill-in all mesh holes. The final models should have similar complexity and quality as those created in the first approach.

Thanks to the detailed labor reporting, it was possible to compare the average effort needed to create usable 3D interior models using these two approaches. Unfortunately, Photoscan could not deal to successful recover one room, so finally five rooms could be used for a comparison. The received results indicate that the approach based on 3D photogrammetry scanning was faster by nearly 30%, in the average (Table 1). Closer analysis shows that the advantage appears mainly for larger and more complex interiors. It is a positive fact as just such rooms consume most modeling time, what is confirmed by the weighted average profit of 41%. This rough comparison confirmed the thesis underlying the sense of the whole project. Moreover, the advantage of the proposed approach would be much higher and reach even 50%, if there existed tools aiding the 3D scan improving and post-processing. Such tools will be developed in the further stage of the STERIO project.

Table 1. The comparison of modeling time for the traditional and photogrammetry approach. Time given in hours [h]

	Room3	Room4	Room1	SmallRoom	Parking	Totally
Traditional approach [h]	180	103	95	18.5	13	409.5
Photogrammetry approach [h]	118	60	82.5	19	12	291.5
Time advantage	34.4%	44.8%	13.1%	−2.6%	7.7%	28.8%

4.2   Comparison of Different Patterns Types for Reconstruction of Flat Surfaces

The goal of this experiment was to compare and verify different types and classes of patterns that can be used to improve reconstruction of interiors using the proposed stereo-photogrammetric approach. In general, several classes of patterns have been identified, including photographs, semi-regular geometric motifs and pseudo-random patterns. For each class several sample images were chosen (Fig. 3). For the experiment a stereo-photography kit were used equipped with a pair of high-resolution Sony A7R cameras. The experiment was carried out for two single-colored walls, one of which ended with a smooth, cylindrical column.

The experiment results were evaluated manually by a human expert. Such approach was sufficient, because the aim of the experiment was to estimate the suitability of individual classes of pattern, and not to accurately determine the errors of reconstruction of scanned surfaces. Analyzing the reconstruction results, one can draw the following conclusions:

– Regular geometric patterns (Fig. 3a–b) do not allow to reconstruct the whole flat surfaces as particular landmarks are too similar one to each other and

may be easily mistaken. Additionally, they add more or less regular moiré effect displacements that must be filtered out at post processing stage.
– Adding some color to pattern variations (Fig. 3c–d) allows to reconstruct the whole flat surfaces in most cases but still does not allow for reconstruction of the curved surfaces. The moiré effect is still clearly visible.
– Pseudo-random motifs (Fig. 3e–f) allow to reconstruct all surfaces still limiting the local displacements which has more random character that is easier to filter out.

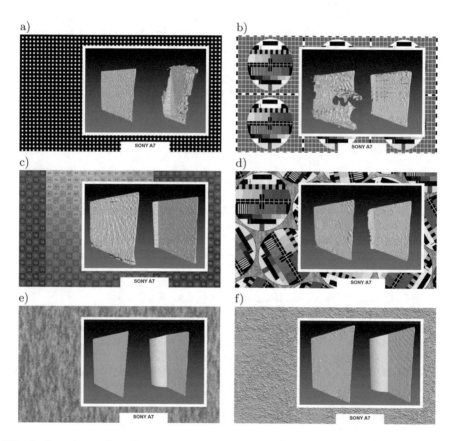

Fig. 3. Sample results of geometry reconstruction of uniform surfaces (a wall and a column) using different displayed pattern classes

5   Conclusion and Future Work

In this paper, a new approach was proposed to speed-up the reconstruction of real interiors for video game purposes. The unique experiment confirmed that using photogrammetric approach can partially automatize this important element of 3D game production pipeline and shorten the modeling time of more complex interiors. Although in many cases, a collection of individual photographs (photogrammetry) may be sufficient to reconstruct a room, serious problems with the reconstruction of flat uniform surfaces often arise. That is why, within the STERIO we propose the stereo-photogrammetric approach. The conducted experiment proved that proposed novel method is fully useful for scanning such problematic surfaces.

Future work includes further development of stereo-photogrammetry methods and a set of tools to maximize the automation level of the entire process with the options of manually editing and performing advanced post-processing algorithms. These tools will be verified and validated within the production of a commercial 3D horror video game, which action will take place in the interiors of an abandoned manor house modeled on existing historic building.

Acknowledgment. This work was supported by the Sectoral Programme GAMEINN within the Operational Programme Smart Growth 2014-2020 under the contract no POIR.01.02.00-00-0140/16.

References

1. Agisoft, L.: Agisoft photoscan user manual: professional edition (2014)
2. Ahmadabadian, A.H., Robson, S., Boehm, J., Shortis, M., Wenzel, K., Fritsch, D.: A comparison of dense matching algorithms for scaled surface reconstruction using stereo camera rigs. ISPRS J. Photogramm. Remote. Sens. **78**, 157–167 (2013)
3. Balsa-Barreiro, J., Fritsch, D.: Generation of visually aesthetic and detailed 3D models of historical cities by using laser scanning and digital photogrammetry. Digit. Appl. Archaeol. Cult. Herit. **8**, 57–64 (2018)
4. Baya, H., Essa, A., Tuytelaarsb, T., Van Goola, L.: Speeded-up robust features (surf). Comput. Vis. Image Underst. **110**(3), 346–359 (2008)
5. Escorcia, V., Dávila, M.A., Golparvar-Fard, M., Niebles, J.C.: Automated vision-based recognition of construction worker actions for building interior construction operations using RGBD cameras. In: Construction Research Congress 2012: Construction Challenges in a Flat World, pp. 879–888 (2012)
6. Ha, H., Oh, T.H., Kweon, I.S.: A multi-view structured-light system for highly accurate 3D modeling. In: 2015 International Conference on 3D Vision (3DV), pp. 118–126. IEEE (2015)
7. Kaczmarek, A.L.: Improving depth maps of plants by using a set of five cameras. J. Electron. Imaging **24**(2), 023018 (2015)
8. Kaczmarek, A.L.: Stereo vision with equal baseline multiple camera set (EBMCS) for obtaining depth maps of plants. Comput. Electron. Agric. **135**, 23–37 (2017)
9. Kazmi, W., Foix, S., Alenya, G.: Plant leaf imaging using time of flight camera under sunlight, shadow and room conditions. In: 2012 IEEE International Symposium on Robotic and Sensors Environments (ROSE), pp. 192–197. IEEE (2012)
10. Khoshelham, K., Elberink, S.O.: Accuracy and resolution of kinect depth data for indoor mapping applications. Sensors **12**(2), 1437–1454 (2012)
11. Rebecca, O., Gold, C., Kidner, D.: 3D city modelling from LIDAR data. In: Advances in 3D geoinformation systems, pp. 161–175. Springer, Heidelberg (2008)

12. Szwoch, M., Kaczmarek, A.L., Dariusz, B.: STERIO - reconstruction of 3D scenery for video games using stereo-photogrammetry. Computer Game Innovations in Monograph of the Lodz University of Technology, Lodz University of Technology, Lodz, Poland, napieralski p, wojciechowski a edn. (2017)
13. Szwoch, M., Pieniażek, P.: Detection of face position and orientation using depth data. In: Image Processing and Communications Challenges 7, pp. 239–251. Springer, Cham (2016)

Integration of Linear SVM Classifiers in Geometric Space Using the Median

Robert Burduk$^{(\boxtimes)}$ and Jędrzej Biedrzycki

Department of Systems and Computer Networks, Wroclaw University of Science
and Technology, Wybrzeze Wyspianskiego 27, 50-370 Wroclaw, Poland
robert.burduk@pwr.edu.pl

Abstract. An ensemble of classifiers can improve the performance of a
pattern recognition system. The task of constructing multiple classifier
systems can be generally divided into three steps: generation, selection
and integration. In this paper, we propose an integration process which
takes place in the geometric space. It means that the fusion of base
classifiers is done using decision boundaries. In our approach, we use the
linear SVM model as a base classifier, the selection process is based on
the accuracy and the final decision boundary is calculated by using the
median of the decision boundary. The aim of the experiments was to
compare the proposed algorithm and the majority voting method.

Keywords: Ensemble of classifiers · Multiple classifier system
Decision boundary · Linear SVM

1 Introduction

The problem of using simultaneously multiple base classifiers for making decisions as to the membership of an object in a class label has been discussed in
works associated with the classification systems for about twenty years [7,18].
The systems consisting of more than one base classifier are called ensembles of
classifiers (EoC) or multiple classifiers systems (MCSs) [5,8,10,17]. The reasons
for the use of a classifier ensemble include, for example, the fact that single classifiers are often unstable (small changes in input data may result in creation of
very different decision boundaries).

The task of constructing MCSs can be generally divided into three steps:
generation, selection and integration [1]. In the first step a set of base classifiers is trained. There are two ways, in which base classifiers can be trained. The
classifiers, which are called homogeneous are of the same type. However, random-
ness is introduced to the learning algorithms by initializing training objects with
different weights, manipulating the training objects or using different features
subspaces. The classifiers, which are called heterogeneous, belong to different
machine learning algorithms, but they are trained on the same data set. In this
paper, we will focus on homogeneous classifiers which are obtained by applying
the same classification algorithm to different learning sets.

© Springer Nature Switzerland AG 2019
M. Choraś and R. S. Choraś (Eds.): IP&C 2018, AISC 892, pp. 30–36, 2019.
https://doi.org/10.1007/978-3-030-03658-4_4

The second phase of building MCSs is related to the choice of a set of classifiers or one classifier from the whole available pool of base classifiers. If we choose one classifier, this process will be called the classifier selection. But if we choose a subset of base classifiers from the pool, it will be called the ensemble selection. Generally, in the ensemble selection, there are two approaches: the static ensemble selection and the dynamic ensemble selection [1]. In the static classifier selection one set of classifiers is selected to create EoC during the training phase. This EoC is used in the classification of all the objects from the test set. The main problem in this case is to find a pertinent objective function for selecting the classifiers. Usually, the feature space in this selection method is divided into different disjunctive regions of competence and for each of them a different classifier selected from the pool is determined. In the dynamic classifier selection, also called instance-based, a specific subset of classifiers is selected for each unknown sample [4]. It means that we are selecting different EoCs for different objects from the testing set. In this type of the classifier selection, the classifier is chosen and assigned to the sample based on different features or different decision regions [6]. The existing methods of the ensemble selection use the validation data set to create the so-called competence region or level of competence.

The integration process is widely discussed in the pattern recognition literature [13,16]. One of the existing ways to categorize the integration process is using the outputs of the base classifiers selected in the previous step. Generally, the output of a base classifier can be divided into three types [11].

- The abstract level – the classifier ψ assigns the unique label j to a given input x.
- The rank level – in this case for each input (object) x, each classifier produces an integer rank array. Each element within this array corresponds to one of the defined class labels. The array is usually sorted and the label at the top is the first choice.
- The measurement level – the output of a classifier is represented by a confidence value (CV) that addresses the degree of assigning the class label to the given input x. An example of such a representation of the output is a posteriori probability returned by Bayes classifier. Generally, this level can provide richer information than the abstract and rank levels.

In this paper we propose the concept of the classifier integration process which takes place in the geometric space. It means that we use the decision boundary in the integration process. In other words, the fusion of base classifiers is done using decision boundaries. The algorithm presented in the paper concerns the case of a two-dimensional feature space.

The geometric approach discussed in [12] is applied to find characteristic points in the geometric space. These points are then used to determine the decision boundaries. Thus, the results presented in [14] do not concern the process of integration of base classifiers, but a method for creating decision boundaries.

The remainder of this paper is organized as follows. Section 2 presents the basic concept of the classification problem and EoC. Section 3 describes the proposed method for the integration base classifiers in the geometric space in which

the median is used to determine the decision boundary. The experimental evaluation is presented in Sect. 4. The discussion and conclusions from the experiments
are presented in Sect. 5.

2　Basic Concept

Let us consider the binary classification task. It means that we have two class
labels $\Omega = \{0, 1\}$. Each pattern is characterized by the feature vector x. The
recognition algorithm Ψ maps the feature space x to the set of class labels Ω
according to the general formula:

$$\Psi(x) \in \Omega. \tag{1}$$

Let us assume that $k \in \{1, 2, ..., K\}$ different classifiers $\Psi_1, \Psi_2, \ldots, \Psi_K$ are
available to solve the classification task. In MCSs these classifiers are called base
classifiers. In the binary classification task, K is assumed to be an odd number.
As a result of all the classifiers' actions, their K responses are obtained. Usually
all K base classifiers are applied to make the final decision of MCSs. Some
methods select just one base classifier from the ensemble. The output of only
this base classifier is used in the class label prediction for all objects. Another
option is to select a subset of the base classifiers. Then, the combining method
is needed to make the final decision of EoC.

The majority vote is a combining method that works at the abstract level.
This voting method allows counting the base classifiers outputs as a vote for
a class and assigns the input pattern to the class with the majority vote. The
majority voting algorithm is as follows:

$$\Psi_{MV}(x) = \arg\max_{\omega} \sum_{k=1}^{K} I(\Psi_k(x), \omega), \tag{2}$$

where $I(\cdot)$ is the indicator function with the value 1 in the case of the correct
classification of the object described by the feature vector x, i.e. when $\Psi_k(x) = \omega$.
In the majority vote method each of the individual classifiers takes an equal part
in building EoC.

3　Proposed Method

The author's earlier work [2] presents results of the integration base classifier in
the geometric space in which the base classifiers use Fisher's classification rule,
while the process of the base classifier selection is performed in the regions of
competence defined by the intersection points of decision boundary functions.
The paper presents [3] results of the integration base classifier in the geometric
space in which the harmonic mean is used. In this paper we use a different linear
classifier', the competence region is defined by the even division of the feature
space and in addition, the median is used to determine the decision boundary
of EoC.

The calculation of EoC decision boundary is performed as follows.

Step 1: Train each of base classifiers $\Psi_1, \Psi_2, \ldots, \Psi_K$ using different training sets by splitting according to the cross-validation rule.
Step 2: Divide the feature space in different separable regions of competence.
Step 3: Evaluate the base classifiers competence in each region of competence based on the accuracy.
Step 4: Select l best classifiers from all base classifiers for each region of competence, where $1 < l < K$.
Step 5: Define the decision boundary of the proposed EoC classifier Ψ_{MGS} as median of the decision boundary in each decision region separately.

The decision boundary obtained in step 5 is applied to make the final decision of the proposed EoC. The calculation of one decision boundary for several base classifiers is a fusion method which is carried out in the geometric space. For binary data sets the proposed algorithm can be used for an even number of basic classifiers (after selection), in contrast to the MV method.

4   Experimental Studies

The main aim of the experiments was to compare the quality of classifications of the proposed method of integration base classifiers in the geometric space Ψ_{MGS} with the majority voting rule (MV) Ψ_{MV}. In the experiment we use the linear SVM algorithm as a base classifier and we created a pool of classifiers consisting of nine base classifiers. In selection steps we removed sequentially one basic classifier from the pool of classifiers, for example Ψ_{MGS}^8, which means that EoC after the selection consisted of eight base classifiers (one base classifier has been removed).

In the experiment the entire space of features has been divided into three equal regions of competence, for each data set separately. In the experimental research we use 12 publicly available binary data sets from UCI machine learning repository and the KEEL Project. For all data sets the feature selection process [9,15] was performed to indicate two most informative features.

In order to compare the quality of the classification, we used two measures. Table 1 shows the results of the classification accuracy (ACC) and the mean ranks obtained by the Friedman test. Table 2 shows the results of the Matthews correlation coefficient (MCC) and the mean ranks obtained by the Friedman test.

Lower average ranks mean better quality of classification. The obtained results indicate that the selection of base classifiers affects the quality of the classification in the proposed method of integrating base classifiers in the geometric space. In the case of conducted experiments, the best results were obtained by EoC composed of five (measure ACC) or four (measure MCC) base classifiers. The proposed algorithm for integrating base classifiers Ψ_{MGS} revealed better results than the baseline classifier which is Ψ_{MV}. However, the post-hoc Nemenyi test at $p = 0.05$ and $p = 0.1$, is not powerful enough to detect any significant differences between the proposed algorithm and the MV method.

Table 1. Classification accuracy and mean rank positions for the proposed method Ψ_{MGS} and the majority voting method Ψ_{MV} produced by the Friedman test

Data set	Ψ_{MGS}^2	Ψ_{MGS}^3	Ψ_{MGS}^4	Ψ_{MGS}^5	Ψ_{MGS}^6	Ψ_{MGS}^7	Ψ_{MGS}^8	Ψ_{MV}
Biodeg	0.747	0.747	0.748	0.748	0.748	0.748	0.748	0.748
Bupa	0.931	0.937	0.911	0.937	0.922	0.938	0.925	0.938
Cryotherapy	0.888	0.888	0.886	0.887	0.887	0.891	0.889	0.891
Banknote	0.724	0.725	0.724	0.725	0.724	0.725	0.724	0.723
Haberman	0.947	0.948	0.95	0.948	0.948	0.947	0.946	0.952
Ionosphere	0.933	0.933	0.935	0.936	0.939	0.938	0.941	0.94
Meter	0.927	0.926	0.926	0.925	0.924	0.923	0.924	0.923
Pop failures	0.886	0.889	0.893	0.891	0.896	0.889	0.893	0.839
Seismic	0.577	0.582	0.576	0.583	0.581	0.582	0.578	0.582
Twonorm	0.775	0.778	0.775	0.775	0.773	0.772	0.771	0.77
Wdbc	0.825	0.88	0.859	0.889	0.872	0.889	0.877	0.901
Wisconsin	0.81	0.805	0.807	0.799	0.791	0.765	0.755	0.743
Mean Rank	5.29	4.00	4.83	3.58	4.75	4.12	5.00	4.41

Table 2. Matthews correlation coefficient and mean rank positions for the proposed method Ψ_{MGS} and the majority voting method Ψ_{MV} produced by the Friedman test

Data set	Ψ_{MGS}^2	Ψ_{MGS}^3	Ψ_{MGS}^4	Ψ_{MGS}^5	Ψ_{MGS}^6	Ψ_{MGS}^7	Ψ_{MGS}^8	Ψ_{MV}
Biodeg	0	0	0	0	0	0	0	0
Bupa	0.019	−0.003	0.047	−0.002	0.027	0	0.023	0
Cryotherapy	0.775	0.774	0.771	0.772	0.771	0.78	0.776	0.779
Banknote	0.331	0.331	0.326	0.328	0.323	0.327	0.325	0.319
Haberman	0.829	0.836	0.84	0.833	0.835	0.83	0.825	0.894
Ionosphere	0.859	0.858	0.862	0.865	0.87	0.868	0.875	0.872
Meter	0.462	0.448	0.45	0.443	0.444	0.443	0.443	0.435
Pop failures	0.755	0.759	0.767	0.763	0.774	0.757	0.766	0.63
Seismic	0.13	0.149	0.137	0.161	0.161	0.158	0.139	0.108
Twonorm	0.257	0.264	0.24	0.223	0.217	0.203	0.198	0.205
Wdbc	0.664	0.771	0.726	0.792	0.756	0.791	0.769	0.815
Wisconsin	0.631	0.62	0.627	0.606	0.587	0.528	0.508	0.48
Mean rank	4.50	4.16	4.00	4.33	4.20	4.66	4.95	5.16

5   Conclusion

In this paper we have proposed a concept of a classifier integration process taking place in the geometric space. In the presented approach, the final decision boundary is calculated by using the median of decision boundary and is performed after the selection process of the base classifiers. In addition, the space of features is divided into regions of competence. For binary data sets the proposed algorithm can be used for an even number of basic classifiers (after selection), in contrast to the MV method.

The experiments have been carried out on twelve benchmark data sets. The aim of the experiments was to compare the proposed algorithm Ψ_{MGS} and the MV method Ψ_{MV}. The results of the experiment show that the selection process affects the quality of the classification. The best results were achieved when EoC was reduced by about half of the original number of base classifiers. Additionally, the proposed algorithm for integrating base classifiers Ψ_{MGS} revealed better results than the baseline classifier which is Ψ_{MV}. However, the post-hoc statistical test did not show a statistical difference in the results.

Acknowledgments. This work was supported in part by the National Science Centre, Poland under the grant no. 2017/25/B/ST6/01750.

References

1. Britto, A.S., Sabourin, R., Oliveira, L.E.: Dynamic selection of classifiers-a comprehensive review. Pattern Recogn. **47**(11), 3665–3680 (2014)
2. Burduk, R.: Integration base classifiers based on their decision boundary. In: Artificial Intelligence and Soft Computing–ICAISC 2017, vol. 10246. LNCS, pp. 13–20. Springer, Heidelberg (2017)
3. Burduk, R.: Integration base classifiers in geometry space by harmonic mean. In: Artificial Intelligence and Soft Computing–ICAISC 2018, vol. 10841. LNCS, pp. 585–592. Springer, Heidelberg (2018)
4. Cavalin, P.R., Sabourin, R., Suen, C.Y.: Dynamic selection approaches for multiple classifier systems. Neural Comput. Appl. **22**(3–4), 673–688 (2013)
5. Cyganek, B.: One-class support vector ensembles for image segmentation and classification. J. Math. Imaging Vis. **42**(2–3), 103–117 (2012)
6. Didaci, L., Giacinto, G., Roli, F., Marcialis, G.L.: A study on the performances of dynamic classifier selection based on local accuracy estimation. Pattern Recogn. **38**, 2188–2191 (2005)
7. Drucker, H., Cortes, C., Jackel, L.D., LeCun, Y., Vapnik, V.: Boosting and other ensemble methods. Neural Comput. **6**(6), 1289–1301 (1994)
8. Giacinto, G., Roli, F.: An approach to the automatic design of multiple classifier systems. Pattern Recogn. Lett. **22**, 25–33 (2001)
9. Guyon, I., Elisseeff, A.: An introduction to variable and feature selection. J. Mach. Learn. Res. **3**, 1157–1182 (2003)
10. Korytkowski, M., Rutkowski, L., Scherer, R.: From ensemble of fuzzy classifiers to single fuzzy rule base classifier. In: Artificial Intelligence and Soft Computing–ICAISC 2008, vol. 5097. LNCS, pp. 265–272. Springer, Heidelberg (2008)
11. Kuncheva, L.I.: Combining Pattern Classifiers: Methods and Algorithms. Wiley, New York (2004)
12. Li, Y., Meng, D., Gui, Z.: Random optimized geometric ensembles. Neurocomputing **94**, 159–163 (2012)
13. Ponti Jr., M.P.: Combining classifiers: from the creation of ensembles to the decision fusion. In: 2011 24th SIBGRAPI Conference on Graphics, Patterns and Images Tutorials (SIBGRAPI-T), pp. 1–10. IEEE (2011)

14. Pujol, O., Masip, D.: Geometry-based ensembles: toward a structural characterization of the classification boundary. IEEE Trans. Pattern Anal. Mach. Intell. **31**(6), 1140–1146 (2009)
15. Rejer, I.: Genetic algorithms for feature selection for brain computer interface. Int. J. Pattern Recognit. Artif. Intell. **29**(5), 1559008 (2015)
16. Tulyakov, S., Jaeger, S., Govindaraju, V., Doermann, D.: Review of classifier combination methods. In: Machine Learning in Document Analysis and Recognition, pp. 361–386. Springer, Heidelberg (2008)
17. Woźniak, M., Graña, M., Corchado, E.: A survey of multiple classifier systems as hybrid systems. Inf. Fusion **16**, 3–17 (2014)
18. Xu, L., Krzyzak, A., Suen, C.Y.: Methods of combining multiple classifiers and their applications to handwriting recognition. IEEE Trans. Syst. Man Cybern. **22**(3), 418–435 (1992)

Reliability of Local Ground Truth Data for Image Quality Metric Assessment

Rafał Piórkowski$^{(\boxtimes)}$ and Radosław Mantiuk

West Pomeranian University of Technology, Szczecin, Poland
{rpiorkowski,rmantiuk}@zut.edu.pl

Abstract. Image Quality Metrics (IQMs) automatically detect differences between images. For example, they can be used to find aliasing artifact in the computer generated images. An obvious application is to test if the costly anti-aliasing techniques must be applied so that the aliasing is not visible to humans. The performance of IQMs must be tested based on the ground truth data, which is a set of maps that indicate the location of artifacts in the image. These maps are manually created by people during so called marking experiments. In this work, we evaluate two different techniques of marking. In the side-by-side experiment, people mark differences between two images displayed side-by-side on the screen. In the flickering experiment, images are displayed at the same location but are exchanged over time. We assess the performance of each technique and use the generated reference maps to evaluate the performance of the selected IQMs. The results reveal the better accuracy of the flickering technique.

Keywords: Aliasing · Image quality metrics
Perceptual experiments · Ground truth data

1 Introduction

Full reference *Image Quality Metrics* (IQMs) assess the visual difference between two images [5,12,13]. This functionality is useful in many visual computing applications. IQMs have been successfully used to automatically evaluate the quality of the realistic computer graphics [2,3], to find artifacts in game engines [9] and artifacts in images reconstructed from raw sensor data [11]. IQMs were also used to assess the quality of 3D mesh after simplification [4] or after replacing triangles with texture [10].

The goal of IQMs is to find differences between images that are also noticeable for a human observer. IQMs subgroup - called *visibility metrics* - generates *difference map*, which points out the local distribution of the differences. To test the correctness of an IQM the reference data collected during a perceptual experiment with human observers is needed. People manually mark differences that they can notice in images. This experimental methodology has a few important disadvantages that lead to the biased and incorrect creation of the reference

© Springer Nature Switzerland AG 2019
M. Choraś and R. S. Choraś (Eds.): IP&C 2018, AISC 892, pp. 37–45, 2019.
https://doi.org/10.1007/978-3-030-03658-4_5

difference maps (consisting of human markings). Marking experiments are time-consuming. Longer time usually leads to more complete and precise marking, however, inter-subject differences are common because people approach the task with various diligence. Also, marking with pixel precision is difficult to achieve in practice. The even more important issue is a *visual search* problem. People often do not mark strong differences, because these differences are masked by the content of the image.

In this work, we analyze how to achieve the valuable reference difference maps based on the marking experiment. We compare two types of experiments: *side-by-side*, in which images are presented side-by-side and observers mark differences in one image, and *flickering*, in which images are displayed at the same location and replaced over time. The resulting difference maps are binarized with 30%, 50%, and 80% thresholds, which means that a pixel is assumed to be marked as it was marked by an arbitrary percentage of observers. Additionally, following Wolski et al. [15], we model the distribution of the visual search and compute a probability of noticing a particular difference while observed by a given number of people. in this analysis, the reference difference map is weighted by an uncertainty of the visual search.

In our study, we concentrate on the *aliasing* [1, Sect. 5.6]. Computer generated images free from the aliasing artifacts are compared to images with aliasing but generated using the multi-sampling anti-aliasing technique. The goal is to show if IQMs are able to objectively evaluate the effectiveness of the anti-aliasing techniques, i.g if they able to mask the aliasing artifacts.

Section 2 presents details of both performed perceptual experiments. In Sect. 3 we present our approach to model reliability ground truth data. This approach is tested and compared with the typical techniques in Sect. 4.

2   Experimental Evaluation

In this section, we discuss the process of collecting ground truth data during perceptual experiments based on two different methodologies: side-by-side and flickering.

2.1   Stimuli

We chose three graphics engines that deliver development environments for the independent developers: Unity 3d (see http://www.unity3d.com), CryEngine 3 (see http://www.cryengine.com), and Unreal Engine 4 (see http://www.unrealengine.com). In these engines one can model a scene using external graphics objects and/or some example scenes delivered with the engine. We modelled 16 scenes that were used to render 36 pairs of test and reference images. Each pair consists of the test image with a varied strength of aliasing artifacts (no anti-aliasing, 2x or 4x MSAA) and the reference image rendered using the most effective anti-aliasing technique (8x or better MSAA). The images were captured in a resolution of 800 × 600 pixels using FRAPS application. We avoided any camera movement between a reference and test images to make the scenes completely static.

2.2   Side-by-Side Vs. Flickering

We asked people to manually mark visible differences between the reference image and an image with a particular artifact. Observers used a brush-paint interface controlled by the computer mouse. The brush size could be reduced up to per-pixel resolution. This procedure was repeated for every pair of images, resulting in 720 comparisons among both experimental methodologies.

The experiment was repeated twice using side-by-side and flickering methodologies. In the side-by-side technique, both images were displayed next to each other. In this approach, observer shift gaze from one image to another looking for artifacts that should be marked in the test image. On the other hand, in the flickering technique, test and reference images were displayed alternately at the same location with the frequency of one second. Differences between images were visible as flickering in various image locations.

2.3   Participants and Apparatus

We repeated markings 10 times for each pair of images and for both experiment types. Finally, we collected 360 difference maps for each type of experiment. To avoid fatigue with an experiment which could result in a decrease in the accuracy of marking, observers were asked to mark only three pairs of images during one experimental session. The remaining images were marked on the following days but an observer could take park only in one type of the experiment. In total, 55 observers performed the side-by-side approach and 60 observers the flickering approach (97 males and 18 females). Observers age ranged from 20 to 47 years.

The experiments were conducted using 24' Eizo ColorEdge CG245W display with a native resolution of 1920×1200 pixels. The display was calibrated to sRGB color profile with the maximum luminance of $110 \, \text{cd/m}^2$. We asked participants to keep distance between their eyes and the display at about 70 cm. Experiments were performed in a darkened room in order to avoid highlights from the surrounding lights.

2.4   Results

Figure 1 presents an example difference map with the markings. Image on the left shows a map created by an individual observer, while the image on the right is a map averaged over 10 observers. It is worth noting that structure of marking for the side-by-side and flickering approach is different. In the flickering experiments, observers tend to mark more areas and agreement between observers was higher.

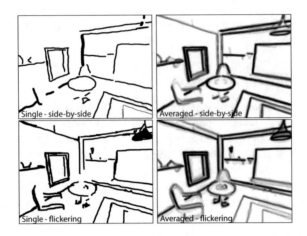

Fig. 1. Left: reference maps created by individual observer. Right: maps averaged over all observers

3   Modeling Reliability of Experimental Data

As can be seen in Fig. 2 (top), the difference maps for the same pair of images but marked by different observers can differ significantly. Moreover, some artifacts are often skipped because people did not notice them despite the large difference in colors (see Fig. 2, bottom). This artifact would certainly be marked if one would draw observers' attention on them.

The key question is how many of observers should mark particular pixel in order to consider this pixel as being a visible difference. In literature, this threshold is typically set to an arbitrary value of 50% [3,8,11].

To improve the difference maps, we performed statistical analysis of the marking against the actual color difference. We assume that an artifact should be marked by people if:

1. Observer did not make a mistake while marking (e.g. she/he marked also pixels in the surrounding of the artifact).
2. Observer visually searched an area with the artifact.
3. Observer was able to detect the artifact.

The probability P of marking the artifact by k out of N observers can be described by the Bernoulli distribution:

$$\begin{aligned} P &= p_{mis} + (1 - p_{mis})\binom{N}{k}(p_{att}{\cdot}p_{det})^k (1 - p_{att}{\cdot}p_{det})^{n-k} \\ &= p_{mis} + (1 - p_{mis})\, Binomial(k, N, p_{att}{\cdot}p_{det}), \end{aligned} \tag{1}$$

where p_{mis} is the probability of mistake, p_{att} - probability of attending the artifact, and p_{det} - probability of detection.

Following Wolski et al. [15], for each image pair we found pixels, for which color difference between a reference and test images is greater than 50/255.

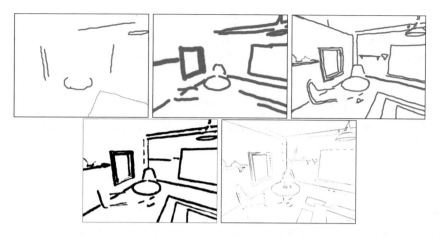

Fig. 2. Top: difference maps created by three different observers for the same pair of the reference and test images. Bottom: difference map averaged over 10 observers and binarized with 50% threshold (left), location of the pixels corresponding to the color difference between reference and test images higher that 50/255 (right)

For these pixels $p_{det} = 1$, so the distribution of p_{att} can be modeled. Figure 3 presents cumulative binomial distribution of p_{att}. The plot shows that for the flickering technique observers marked proper pixels with higher probability comparing to the side-by-side technique. In the case of the side-by-side method, it is apparent that maximum of the probability has its peak at 0.76 and then quickly diminish. In contrary, for the flickering this value is equal to 0.92 and falls to 0.13 for $p = 1$. These two curves suggest that for the side-by-side experiment even high color difference does not guarantee that the artifact will be marked.

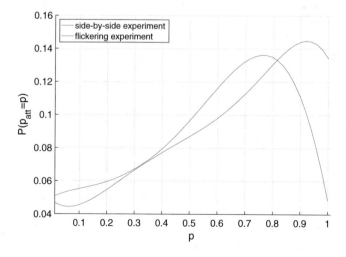

Fig. 3. The probability that the probability of attending a difference is equal to p, plotted separately for side-by-side and flickering experiments

The distribution of p_{att} can be further used to compute the log-likelihood that difference map generated by an IQM correctly predicts the reference data. This likelihood L is expressed by the following equation:

$$L = \sum_{(x,y)\in\Theta} \log[p_{mis} + (1 - p_{mis})$$
$$\cdot \int_0^1 p_{att}(p)\cdot Binomial\,(k(x,y), N, p_{att}(p)\cdot p_{det}(x,y))\;dp], \tag{2}$$

where Θ is a set of all pixels in the image. The details on modeling this likelihood can be found in [15].

4   IQMs Evaluation

4.1   Difference Maps

Image Quality Metrics takes as input two images. First of them is called test image and it is in some way distorted (includes artifacts). Second image - the reference image is considered as the ground truth. It is free from any kind of distortion [12]. As a result of comparison, the IQMs produce greyscale map, called difference map, indicating local differences between input images. The difference map approximates the probability of visibility differences between test and reference map by the average human observer. In order to choose which IQMs is the most effective, one needs to compare the difference maps generated by a given IQM with ground truth data (i.e. the difference map from the experiment described in Sect. 2).

4.2   Evaluation Procedure

In Sect. 2 two techniques of marking local differences in images with- and without aliasing artifacts have been introduced. We use the captured data to evaluate performance of selected IQMs: MSE (Mean Squared Error), SSIM (Structural SIMilarity Index [12]), MS-SSIM (Multiscale Structural SIMilarity Index) [14], CID (Color-Image Difference) [6], S-CIELAB [16], and HDR-VDP-2 [7]. Example difference maps generated by the listed metrics are presented in Fig. 4.

Typical evaluation of a metric performance is based on plotting the Receiver-Operator Curve (ROC) and computing the Area-Under-Curve (AUC) [2,3,9]. AUC expresses metric correctness as one scalar value ranging from 0 to 1, while 1 means a perfect agreement between reference difference maps (human markings) and the difference map generated by a metric.

The results of metric assessment for our dataset are presented in Table 1 and Fig. 5. ROC analysis does not reveal any significant differences between the ground truth data captured using side-by-side or flickering techniques (see Fig. 5 (right)). However, the likelihood for side-by-side is higher than for flickering (see Fig. 5 (left)). It can be interpreted that side-by-side technique is less restrictive and gives the metrics better ratings. Flickering reveals lower metrics performance, however, the overall ranking of the metrics is not changed. i.e. the HDR-VDP-2 is the best metric for detecting aliasing artifacts, followed by SSIM, CID, MS-SSIM, S-CIELAB, and MSE as the worst metric.

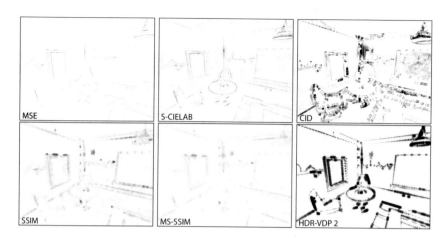

Fig. 4. Examples of the difference maps generated by IQMs

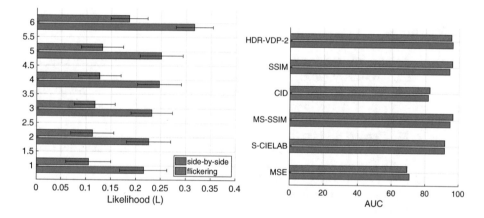

Fig. 5. Likelihood (left) and AUC 50% (right) for side-by-side and flickering experiments. The black lines denote the standard error of mean

Table 1. Average AUC and likelihood values for both experimental techniques

	Side-by-side				Flickering			
	AUC			Likelihood	Likelihood	AUC		
	30%	50%	80%			30%	50%	80%
MSE	67.41	70.52	76.90	0.2157	0.1053	68.45	69.19	74.50
S-CIELAB	89.70	91.60	94.56	0.2258	0.1130	90.16	91.75	94.24
SSIM	94.49	94.41	94.60	0.2504	0.1322	95.80	96.08	96.47
MS-SSIM	94.71	94.84	95.33	0.2316	0.1176	96.16	96.52	97.01
CID	79.40	81.75	84.11	0.2466	0.1269	79.20	82.66	85.51
HDR-VDP 2	95.61	96.21	96.81	0.3165	0.1856	95.03	95.39	96.25

5   Conclusions and Future Work

We compared two experimental methodologies that are used to capture ground truth data for local image differences. The flickering technique revealed better performance from the perspective of the visual search task. The likelihood of detection of data captured using this technique has lower absolute values than for the side-by-side technique. However, the ranking of the metrics is the same for both techniques, which suggests that both techniques deliver good quality reference data.

In future work, we plan to perform a similar comparison for other types of artifacts like Z-fighting, shadow acne, or peter panning. It would be also interesting to look for another technique of manual marking, which would be less time consuming and would deliver valuable ground truth data.

Acknowledgments. The project was partially funded by the Polish National Science Centre (decision number DEC-2013/09/B/ST6/02270).

References

1. Akenine-Möller, T., Haines, E., Hoffman, N.: Real-Time Rendering, 3rd edn. A K Peters Ltd., Wallesley (2008)
2. Čadík, M., Herzog, R., Mantiuk, R., Mantiuk, R., Myszkowski, K., Seidel, H.P.: Learning to predict localized distortions in rendered images. Comput. Graph. Forum **32**(7), 401–410 (2013)
3. Čadík, M., Herzog, R., Mantiuk, R., Myszkowski, K., Seidel, H.P.: New measurements reveal weaknesses of image quality metrics in evaluating graphics artifacts. ACM Trans. Graph. (TOG) **31**(6), 147 (2012)
4. Corsini, M., Larabi, M.C., Lavoué, G., Petřík, O., Váša, L., Wang, K.: Perceptual metrics for static and dynamic triangle meshes. Comput. Graph. Forum **32**(1), 101–125 (2013)
5. Lavoué, G., Mantiuk, R.: Quality assessment in computer graphics. In: Visual Signal Quality Assessment, pp. 243–286. Springer, Cham (2015)
6. Lissner, I., Preiss, J., Urban, P., Lichtenauer, M.S., Zolliker, P.: Image-difference prediction: from grayscale to color. IEEE Trans. Image Process. **22**(2), 435–446 (2013)
7. Mantiuk, R., Kim, K.J., Rempel, A.G., Heidrich, W.: HDR-VDP-2: a calibrated visual metric for visibility and quality predictions in all luminance conditions. ACM Trans. Graph. **30**(4), 40:1–40:14 (2011)
8. Piórkowski, R., Mantiuk, R.: Using full reference image quality metrics to detect game engine artefacts. In: Proceedings of the ACM SIGGRAPH Symposium on Applied Perception, pp. 83–90. ACM (2015)
9. Piórkowski, R., Mantiuk, R., Siekawa, A.: Automatic detection of game engine artifacts using full reference image quality metrics. ACM Trans. Appl. Percept. (TAP) **14**(3), 14 (2017)
10. Rushmeier, H.E., Rogowitz, B.E., Piatko, C.: Perceptual issues in substituting texture for geometry. In: Electronic Imaging, pp. 372–383. International Society for Optics and Photonics (2000)

11. Sergej, T., Mantiuk, R.: Perceptual evaluation of demosaicing artefacts. In: Image Analysis and Recognition. LNCS, vol. 8814, pp. 38–45. Springer, Cham (2014)
12. Wang, Z., Bovik, A., Sheikh, H., Simoncelli, E.: Image quality assessment: from error visibility to structural similarity. IEEE Trans. Image Process. **13**(4), 600–612 (2004)
13. Wang, Z., Bovik, A.: Modern Image Quality Assessment. Morgan & Claypool Publishers (2006)
14. Wang, Z., Simoncelli, E.P., Bovik, A.C.: Multiscale structural similarity for image quality assessment. In: Conference Record of the Thirty-Seventh Asilomar Conference on Signals, Systems and Computers 2004, vol. 2, pp. 1398–1402. IEEE (2003)
15. Wolski, K., Giunchi, D., Ye, N., Didyk, P., Mantiuk, R., Seidel, H.P., Steed, A., Mantiuk, R.K.: Dataset and metrics for predicting local visible differences. ACM Trans. Graph. (2018)
16. Zhang, X., Wandell, B.A.: A spatial extension of cielab for digital color-image reproduction. Journal of the Society for Information Display **5**(1), 61–63 (1997)

Video Processing and Analysis for Endoscopy-Based Internal Pipeline Inspection

Nafaa Nacereddine[1(✉)], Aissa Boulmerka[2], and Nadia Mhamda[1]

[1] Research Center in Industrial Technologies CRTI, 16014 Algiers, Algeria
{n.nacereddine,n.mhamda}@crti.dz
[2] DMI, Centre Universitaire de Mila, 43000 Mila, Algeria
a.boulmerka@centre-univ-mila.dz

Abstract. Because of the increasing requirements in regards to the pipeline transport regulations, the operators take care to the rigorous application of checking routines that ensure nonoccurrence of leaks and failures. In situ pipe inspection systems such as endoscopy, remains a reliable mean to diagnose possible abnormalities in the interior of a pipe such as corrosion. Through digital video processing, the acquired videos and images are analyzed and interpreted to detect the damaged and the risky pipeline areas. Thus, the objective of this work is to bring a powerful analysis tool for a rigorous pipeline inspection through the implementation of specific algorithms dedicated to this application for a precise delimitation of the defective zones and a reliable interpretation of the defect implicated, in spite of the drastic conditions inherent to the evolution of the endoscope inside the pipeline and the quality of the acquired images and videos.

1 Introduction

The pipelines are most likely to be attacked by corrosion, cracking or manufacturing defect, resulting sometimes in catastrophic damage (human damage, environmental pollution, additional repair costs, extended pumping shutdown, etc.). Consequently, the operators ensure rigorous monitoring schedule to prevent occurrence of any breakdown due to fluid leaks. Nowadays, such a situation poses increasingly severe requirements in terms of rules and standards governing the fluid pipe transport. For the purpose mentioned above, in situ pipe inspection systems such as endoscopy, remains a reliable mean to diagnose possible abnormalities in the interior of a pipe such as corrosion. The latter is defined as the breaking down or destruction of material, especially a metal through oxidation or chemical reactions [1]. The corrosion attacks remain the principal cause of fluid leaks and pipeline breaks.

In this work, the endoscope consists in a motorized engine equipped with a digital camera. The recovered videos and images inside the inspected pipelines

© Springer Nature Switzerland AG 2019
M. Choraś and R. S. Choraś (Eds.): IP&C 2018, AISC 892, pp. 46–54, 2019.
https://doi.org/10.1007/978-3-030-03658-4_6

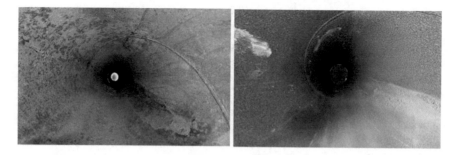

Fig. 1. Examples of corrosion in water transport pipelines

are analyzed and interpreted using digital video processing to detect the damaged and risky areas such as cracks, corrosion, lack of thickness, etc. To this end, the main stages of this work consist in: (1) designing and conceiving an automated motorized engine with embedded camera where the control is insured by Field Programmable Gate Arrays (FPGA) technology, and (2) implementing image/video processing and analysis software for an accurate detection and reliable identification and interpretation of the defective or risky areas such corrosion, found inside the pipeline, as illustrated in Fig. 1.

2 Motorized Vehicle Setup

In case of large pipeline, a vehicle controlled remotely, equipped with a camera, remains a practical solution, timely and interesting in order to perform internal inspection. The automated engine permits to provide real-time feedbacks to the user in terms of internal pipe video sequences to be analysed, distance travelled inside the pipeline, time spent, etc. In this work, to control the engine and the camera, we opt for the FPGA technology since, unlike hard-wired printed circuit board designs, these circuits provide several advantages such as hardware parallelism, flexibility, rapid prototyping capabilities, short time to market, reliability, long-term maintenance, etc. [2]. The realized pipeline endoscope is called "Pipe Explorer" where, the stepper motors driving the wheels are controlled by Zedboard (Xilinx Zynq-7000) FPGA card while the embedded CCD camera acquiring the videos is controlled by MicroZed FPGA card. "Pipe Explorer" is composed of:

- 4 wheels
- 4 stepper motors (1.8° /step) rotating on 360°
- 4 drivers microstep (JK 1545), DC power input type: 24 V 50VDC. Output current: 1.3 A–4.5 A
- 1 EPL camera USB with 2 megapixel of resolution
- Lighting sources through 12 LEDs (12 V) Samsung
- 3 Lead batteries rechargeable of capacity 12 V/9 Ah each, for the motors supply

– 2 Lithium batteries of capacity $12\,V/11\,Ah$ each, to supply Zedboard and MicroZed cards
– 1 Lithium battery $12\,v/11\,A$ to supply the LEDs

For the endoscope thus realized, as shown in Fig. 2, the following characteristics are noted:

– Dimensions: $30cm \times 33\,cm \times 54\,cm$
– Weight: $25\,kg$
– Autonomy: $6\,h$ for a run of $2\,km$.

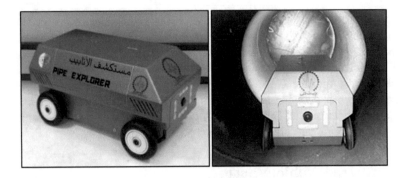

Fig. 2. The endoscope "Pipe-Explorer"

3 Video Analysis Software

In the context of pipeline inspection by endoscopy, we have developed software including video processing and analysis techniques in order to interpret the acquired videos. It is a question here of videos acquired inside the water transportation pipelines which, after a certain time of service, are prone to different types of attacks and degradations, particularly, corrosion. It is thus important to determine its localization, its extent and its degree of severity in order to take appropriate corrective action. Since the corroded part changes its appearance with respect to the rest of the pipeline, the frames or images forming the video sequence must be exploited using image segmentation techniques in order to extract the object representing corrosion, i.e. the defective area, from the background representing the undamaged region of the pipe. In fact, the segmentation constitutes one of the most significant problems in the image analysis system, because the result obtained at the end of this stage strongly governs the final quality of interpretation [3]. Color is a distinctive parameter that can be used to extract the corroded area from the rest of the internal pipeline view. Since these color images can be converted into grayscale images, thresholding techniques become a strong candidate for efficient segmentation. Thresholding is the process of partitioning pixels in the images into object and background classes based upon the relationship between the gray level value of a pixel and a parameter called the threshold. Because of their efficiency in performance and their simplicity in theory, thresholding techniques have been studied extensively and a large number of thresholding methods have been published [4,5]. Generally, for the images obtained from the video sequence acquired inside the pipeline, the overlapping between the corroded region and the background representing the healthy areas is large. This is due to the luminance variability of the corrosion and the background areas, in addition of the presence of artifacts. In such case, by a global thresholding, we do not obtain the desired results. That is why a local adaptive thresholding technique can be employed to overcome the above-mentioned problem.

In locally adaptive thresholding, the threshold value $T(x, y)$ is computed on the neighborhood of the current pixel (x, y), i.e.

$$b(x, y) = \begin{cases} 0 & \text{if } f(x, y) < T(x, y) \\ 1 & \text{otherwise} \end{cases} \tag{1}$$

where, f is the input grayscale image and b is the output binary image.

In order to compute the threshold $T(x, y)$, the local mean $\mu(x, y)$ and the standard deviation $\sigma(x, y)$ have to be computed in a $W \times W$ window centered around each pixel (x, y). In this paper, the method of Sauvola [6], which is an improved version of the method of Niblack [7], and the method of Feng [8], which is an improved version of the method of Wolf [9], are applied to detect in-situ pipeline corrosion and are compared in terms of detection efficiency.

3.1 Sauvola Thresholding Method

The main idea of Niblack's thresholding method [6] is to vary the threshold value, for each pixel (x, y), over the input image of size $(M \times N)$, based on the local mean $\mu(x, y)$ and the local standard deviation $\sigma(x, y)$ in a $W \times W$ window centered around the pixel (x, y). The threshold value at pixel (x, y) is computed by $T(x, y) = \mu(x, y) + k \times \sigma(x, y)$ where, k is a parameter which depends on image content. The parameters W and k are chosen empirically. This method tends to produce a big amount of noise, particularly, when the image background contains light textures, which are considered as object with small contrast. To overcome the above mentioned problems, Sauvola et al. [6] proposed an improved formula to compute the threshold

$$T(x, y) = \mu(x, y) \left(1 - k \left(1 - \frac{\sigma(x, y)}{R} \right) \right) \tag{2}$$

where R is the dynamic range of standard deviation and k a parameter which takes positive values in the range [0.2 0.5]. In this method, the parameter k controls the value of the threshold in the local window such that the higher the value of k, the lower the threshold from the local mean $\mu(x, y)$ [10].

3.2 Feng Thresholding Method

Feng's thresholding method [8] is improved from Wolf et al. [9] thresholding approach in order to tolerate different degrees of illumination unevenness. This method uses two local windows: a primary and a secondary window with the former contained within the latter. The values of local mean $\mu(x,y)$, local standard deviation $\sigma(x,y)$ and minimum gray level M, are calculated in the primary local window and the dynamic range of standard deviation R_σ is calculated in the secondary larger window. The threshold value $T(x,y)$ in a local window is obtained from

$$T(x,y) = (1 - \alpha_1)\mu(x,y) + \alpha_2 \left(\frac{\sigma(x,y)}{R_\sigma} \right) (\mu(x,y) - M) + \alpha_3 M \qquad (3)$$

where $\alpha_2 = k_1 \left(\sigma(x,y)/R_\sigma \right)^\gamma$, $\alpha_3 = k_2 \left(\sigma(x,y)/R_\sigma \right)^\gamma$, and α_1, γ, k_1 and k_2 are positive constants. γ is set to 2 and α_1, k_1 and k_2 are in the ranges of [0.1 0.2], [0.15 0.25] and [0.01 0.05], respectively.

3.3 Algorithms Complexity

Computing $\mu(x,y)$ and $\sigma(x,y)$ in a direct way results in a computational complexity of $O(W^2 M \times N)$ for an $M \times N$ image. In order to speed up the computation, an efficient way of computing local means and variances using sum tables (integral images) is proposed in [11,12] and applied for Sauvola thresholding in [10] so that, the computational complexity does not depend on the window dimension anymore, reducing thus, the computational complexity from $O(W^2 M \times N)$ to $O(M \times N)$. Indeed, for our application which deals with online pipeline inspection, it is important to speed-up the image and video analysis algorithms since slow inspection systems lead to extended pumping downtime where, the cost incurred through lost production would be dramatic. So, the use of integral image in this paper permits to the corrosion detection process based on Sauvola and Feng thresholding methods to be faster since less downtime means more working time and then, more benefits.

4 Experiments

4.1 Software Interface Presentation

We have developed software dedicated to the detection of corrosion where an interactive interface, illustrated in Fig. 3, permits to (1) load the video of our interest, (2) choose the size of the adaptive thresholding window W, (3) choose the values of Sauvola and Feng methods parameters given in Eqs. (2) and (3), namely k, R, α_1, γ, k_1, k_2 and R_σ and (4) finally choose the number of the frames to be processed from the whole video sequence.

Before displaying the segmented frames for both methods, a procedure of processing removal on the area surrounding the disk image, representing the deep

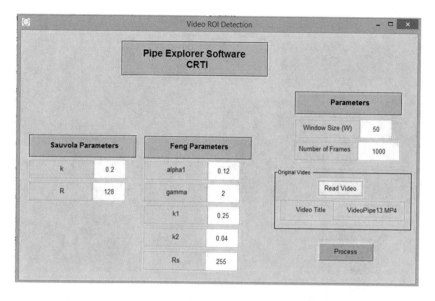

Fig. 3. Presentation of the "Pipe Explorer Software" interface

internal view of the pipeline, is applied (see blue circle in Fig. 4). The processing removal in this area is motivated by the fact that the latter looks very dark since it is very far from the embedded camera and the lighting reaching it is low. Furthermore, the considered regions are gradually processed as the motorized vehicle progresses in the pipeline.

4.2 Region Uniformity Measure

The uniformity of a feature over a region is inversely proportional to the variance of the values of that feature evaluated at every pixel belonging to that region [13]. In this paper, the region uniformity measure U, used to evaluate the segmentation method performance, is given by

$$U = 1 - \frac{w_0\sigma_0^2 + w_1\sigma_1^2}{\sigma_{max}^2} \qquad (4)$$

where (w_0, σ_0) and (w_1, σ_1) are (area ratio, variance) of the foreground and the background regions, respectively; whilst σ_{max}^2 is the maximum image variance given by $(f_{max} - f_{min})^2/2$. The highest (near to 1) is the value of U, the highest is the thresholding quality.

4.3 Results and Discussion

For the test, we have chosen a video sequence obtained by the embedded camera, composed of 1000 frames and stored on a section of water transport pipeline

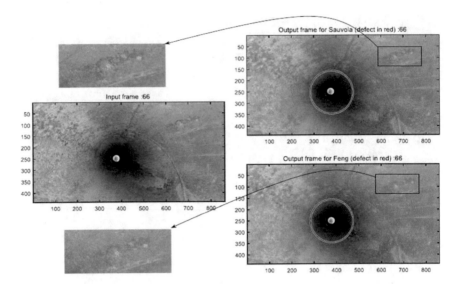

Fig. 4. Sauvola and Feng thresholding results for frame 66

presenting localized attacks of corrosion. The red spots displayed on the output frames representing the risky areas are obtained by Sauvola and Feng thresholding methods. As example, the results for frame 66 are illustrated in Fig. 4. The comparison between the methods of Sauvola and Feng in terms of proportion of damaged areas and thresholding evaluation based region uniformity measure, for all the 1000 images composing the acquired video sequence, are provided by the graphs of Fig. 5a and b, respectively.

It appears from both pictures and graphics that the Sauvola method slightly under-segments the input image compared to the Feng method as shown in Fig. 4 where visually, some parts of the defect indications are missed in the case of Sauvola thresholding. As reported in Table 1, this can be confirmed by the

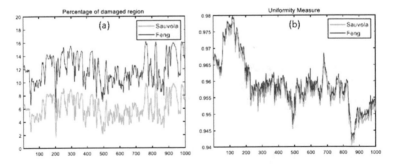

Fig. 5. (a) Percentage of damaged region areas and (b) uniformity measures for methods of Sauvola and Feng on the whole video sequence

Table 1. Sauvola and Feng results for all the 1000 frames

	Sauvola		Feng				
Window size	50		50				
Parameters	k	R	α_1	γ	k_1	k_2	R_s
	0.2	128	0.12	2	0.25	0.04	255
Damaged region							
Range (%)	[1.6% 9.8%]		[5.4% 16.2%]				
Average (%)	6.2%		12%				
Unif. measure							
Range	[0.940 0.978]		[0.942 0.980]				
Average	0.958		0.960%				

measure of the proportions of the damaged region area (colored in red) in regards of the whole processed input image for Sauvola and Feng methods, which are about 5.1% and 10.3%, respectively. The case of frame 66 is valid for all the frames of the video sequence where, the damaged region percentages vary, for Sauvola, in the range of [1.6% 9.8%] with an average of 6.2% whereas, it varies, for Feng, in the range of [5.4% 16.2%] with an average of 12%. Regarding the thresholding performance evaluation, the average values of uniformity measure, on all the video frames, is slightly better for Feng ($U \approx 0.960$) than for Sauvola ($U \approx 0.958$). It is worth to note a possible presence of false positive indications which overestimate the corrosion such as the spiral welding joint in Fig. 4. That said, it is imperative to compare these results with those obtained by an expert for their validation and in order to guide the choice of the thresholding parameters.

5   Conclusion

In this paper, a modern visual equipment "Endoscope" to inspect the interior part of the pipe is used. It consists in a motorized engine embedding a CCD camera. Here, the endoscope is controlled using FPGA technology. A video sequence acquired from a water transport pipeline, presenting mainly corrosion indications, is analyzed and interpreted. Through dedicated software, the indications of corrosion are extracted using two image thresholding methods: Sauvola and Feng. The results are then quantified in terms of area ratio of possible damaged region and in terms of thresholding evaluation measures. To conclude, it reveals from the results that the complex nature of such images requires to the image and video processing techniques to be evaluated in a supervised manner, i.e. by using a ground truth provided by an expert. In other words, a realistic corrosion detection and interpretation should take into account a priori knowledge on the objects composing the internal view of a pipeline such as welding joint, land deposit, stones, stains and soils, which could drastically influence the final interpretation results. These last remarks will be deeply investigated in future works.

Acknowledgments. This research was supported by CRTI as part of technological development of research products with a socio-economic impact. We thank all the team members of project "Pipeline Inspection by Endoscopy" who have participated in the realization of the "Endoscope" which make possible the acquiring of video sequences in pipeline and then their processing: the subject of the study in this paper.

References

1. The free dictionary by Farlex. http://www.thefreedictionary.com/corrosion
2. Yasir, M.: Introduction to FPGA Technology (2011). Website FPGA related.com, https://www.fpgarelated.com/showarticle/17.php
3. Soler, L., Malandrin, G., Delingette, H.: Segmentation automatique: application aux angioscanners 3D du foie. Traitement de Signal **15**(5), 411–431 (1998)
4. Lee, S.U., Chung, S.Y., Park, R.H.: A comparative performance study of several global thresholding techniques for segmentation. Comput. Vis. Graph. Image Process. **52**, 17–190 (1990)
5. Sezgin, M., Sankur, B.: Survey over image thresholding techniques and quantitative performance evaluation. J. Electron. Imaging **13**(1), 146–165 (2004)
6. Sauvola, J., Pietikakinen, M.: Adaptive document image binarization. Pattern Recogn. **33**(2), 225–236 (2000)
7. Niblack, W.: Introduction to Digital Image Processing. Prentice Hall (1986)
8. Feng, M.L., Tan, Y.P.: Contrast adaptive binarization of low quality document images. IEICE Electron. Express **1**(16), 501–506 (2004)
9. Wolf, C., Jolion, J.M.: Extraction and recognition of artificial text in multimedia documents. Pattern Anal. Appl. **6**(4), 309–326 (2004)
10. Shafait, F., Keysers, D., Breuel, T.M.: Efficient implementation of local adaptive thresholding techniques using integral images. In: Document Recognition and Retrieval XV, Proceedings of SPIE, San Jose, CA, vol. 6815 (2008)
11. Crow, F.: Summed-area tables for texture mapping. ACM SIGGRAPH Comput. Graph. **18**, 207–212 (1984)
12. Viola, P., Jones, M.: Rapid object detection using a boosted cascade of simple features. In: Proceedings of IEEE Computer Society Conference on Computer Vision and Pattern Recognition, Kauai, HI, USA, pp. 511–518 (2001)
13. Levine, M.D., Nazif, A.M.: Dynamic Measurement of Computer Generated Image Segmentations. IEEE Transactions on Pattern Analysis and Machine Intelligence **7**(2), 155–164 (1985)

The Influence of Object Refining
in Digital Pathology

Łukasz Roszkowiak[1(✉)], Anna Korzyńska[1], Krzysztof Siemion[1,2],
and Dorota Pijanowska[1]

[1] Nalecz Institute of Biocybernetics and Biomedical Engineering,
Polish Academy of Sciences, Warsaw, Poland
lroszkowiak@ibib.waw.pl
[2] Orlowski Public Clinical Hospital, Center for Medical Postgraduate Education,
231 Czerniakowska Street, 00-416 Warsaw, Poland

Abstract. Quantitative analysis of histopathological sections can be
used to support the diagnosis and evaluate the disease progression by
pathologists. The use of computer-aided diagnosis in pathology can sub-
stantially enhance the efficiency and accuracy of pathologists decisions,
and overall benefit the patient. The evaluation of the shape of specific
types of cell nuclei plays an important role in histopathological examination in various types of cancer. In this study we try to verify how much
the results of segmentation could be improved with applying boundary refinement algorithm to thresholded histopathological image. In this
paper we studied 5 methods based on various approaches: active contour, k-means clustering, and region-growing. For evaluation purposes,
ground truth templates were generated by manual annotation of images.
The performance is evaluated using pixel-wise sensitivity and specificity
metrics. It appears that satisfactory results were achieved only by two
algorithms based on active contour. By applying methodology based on
active contour algorithm we managed to achieve sensitivity of about 93%
and specificity of over 99%. To sum up, thresholding algorithms produce
results that almost never perfectly fit to real object's boundary, but this
initial detection of objects followed by boundary refinement results in
more accurate segmentation.

1 Introduction

The histopathological sections allow pathologists to evaluate a wide range of
specimens obtained from biopsies and surgical procedures. The human direct
evaluation of numerous biopsy slides represents a labor intensive work for pathologists. Quantitative analysis of such sections can be used to support the diagnosis
and evaluate the disease progression. In order to further enhance the efficiency
and accuracy of diagnostic decision, many computer-aided image analysis techniques for the diagnostic images analysis have been proposed [8, 10, 12]. However,
certain obstacles and limitations still exist in achieving a good result, among
which are: variability in shape, size and colour of objects of interest, lack of
defined stain intensity cut-offs as well as overlapping structures.

© Springer Nature Switzerland AG 2019
M. Choraś and R. S. Choraś (Eds.): IP&C 2018, AISC 892, pp. 55–62, 2019.
https://doi.org/10.1007/978-3-030-03658-4_7

The evaluation of the shape of specific types of cell nuclei plays an important role in histopathological examination in various types of cancer [15]. The accuracy of the employed automated cell segmentation technique has major impact in obtaining satisfactory object boundary. After initial thresholding the boundary can be further refined to fit original nuclei more closely.

In this study we try to verify how much the results of preliminary segmentation could be improved with applying boundary refinement algorithm to thresholded histopathological image. To carry out this task we compare different methods of refining the segmented object boundary and recommend a most efficient technique in the case of nuclei segmentation in histopathological images. We validate the efficiency of refinement methods on manually labeled set of images of breast cancer tissue sections immunohistochemically stained against FOXP3 antigen with 3,3'-diaminobenzidine and hematoxylin (DAB&H) to show regulatory T-cell nuclei.

1.1   Related Works

Many different techniques have been proposed for detection and segmentation of cells' nuclei in digital images of tissue sections [8]. Most of the segmentation methods are threshold-based and texture-based and do not take into consideration the boundary of the object explicitly. Alternatively, there are methods that segment images with respect to object boundary such as: geodesic active contour, level-set, clustering. Few other works presented two step approach; detect the objects of interest with thresholding algorithm, and then, refine the shape (boundary) with another algorithm.

Bunyak et al. [5] used similar approach where images first undergo fuzzy c-means segmentation and then multiple algorithms are used for boundary refinement, such as: multiphase vector-based level set, Edge-Based Geodesic Active Contours. The authors reported increase in nuclei segmentation efficiency of about 1 and 3% points respectively. In this case evaluation was based on prostate biopsy histopathology images stained with Hematoxylin&Eosin. Another proposed methodology [1] is based on graph-cut algorithm. First, objects are extracted automatically using a graph-cuts based binarization. Then, a segmentation is refined using a second graph-cuts based algorithm incorporating the method of alpha expansions and graph coloring. Evaluation was based on breast cancer biopsy histopathology images stained with H or DAPI. The authors reported a 6 percent point increase in binarization accuracy.

2   Materials and Methods

2.1   Image Data

In this paper the images of breast cancer patients' tissue sections used for the validation of the experiments were obtained from the Molecular Biology and Research Section, Hospital de Tortosa Verge de la Cinta, Institut d'Investigacio

Sanitaria Pere Virgili (IISPV), URV, Spain, and the Pathology Department of the same hospital. The histological sections were prepared similarly to our previous research [14]. All images show the brown end-products for the immunopositive cells' nuclei among blue colour nuclei for the immunonegative cells as presented in Fig. 1.

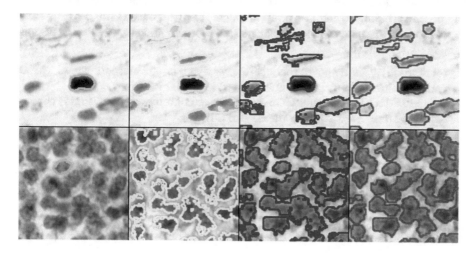

Fig. 1. Two samples from dataset with sparse and compact architecture of cells. RGB image sample with overlaid (from left to right): ground truth example (green); initial thresholding by Bradley method without refinement (yellow); boundary refinement with refine_CV (blue); boundary refinement with refine_AC (red).

To form a dataset for examination 13 randomly selected regions of interest (ROI) of size 1000×1000 pixels were extracted from dataset. The quality of selected image set is typical for this type of biological data. The set consists of images with different degrees of complexity, tissue presented in the images varies from very dark to light and with sparse and compact architecture of cells, as shown in Fig. 1.

2.2   Refinement Methods

Segmentation is generally used to separate objects of interest from the background. In this situation the nuclei are separated from surrounding tissue and empty background. Results of thresholding according to relative color intensity are almost never perfectly fit to real object's boundary. Since, shape of segmented nuclei is an important factor of cells' nuclei detection, we apply refinement methods to achieve a better fit.

In this investigation color deconvolution algorithm was applied to RGB images to separate color information related to H and DAB staining. We obtained monochromatic images representing separate dye layers of H and DAB. Then, we applied 4 methods of adaptive threshold based on intensity: Bradley [3], fcm [2], mce [4] and Nick [9]. Thresholding methods were chosen based on the previous study [14]. This results in nuclei detection with high accuracy but most of the objects are not solid and have areas of incontinuous pixels – as shown in Fig. 1 (second column) and Fig. 2 (first column). The processing goes as follows, every object undergoes morphological operation of filling and closing. Then, the refinement algorithm is applied to each object separately. Moreover, we use simple post-processing to remove small artifacts from the output image.

We tested 5 boundary refinement methods based on: active contours, region-growing and k-means:

- refine_AC [13] implements the localized active contour method using level set method; thresholding results are treated as initial contour;
- refine_CV [16] implements Chan-Vese active contours without edges; thresholding results are treated as initial contour;
- refine_RG [11] implements region-growing from seed point using intensity mean measure; starting point is based on thresholding results after ultimate erosion;
- refine_RRG [6] implements recursive region growing algorithm; starting point is based on thresholding results after ultimate erosion;
- refine_KM [7] implements K-means image segmentation based on histogram; pixels are allocated to two clusters.

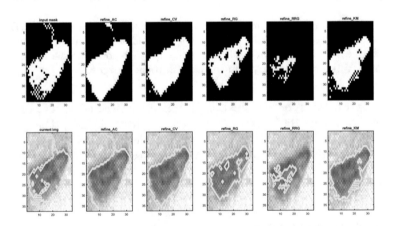

Fig. 2. Comparison of 5 boundary refinement methods results. In first column initial mask (result of thresholding) is presented as binary image and overlaid on grayscale DAB layer. Similarly results of boundary refinement methods are presented in following column in order: refine_AC, refine_CV, refine_RG, refine_RRG, refine_KM.

2.3 Evaluation

The main objective of the evaluation is to determine if the boundaries of segmented objects obtained by the thresholding method and then refined are consistent with the manually labeled ones. As well as to estimate which method gives best refined boundary.

To evaluate and to compare the methods with each other two metrics were used, namely not only the sensitivity and the specificity of object detection but also the erroneously labeled pixels defined as the percentages of incorrectly labeled pixels before and after boundary refinement in relation to ground truth.

In order to evaluate and compare the detection performance of the tested methods, the location of nuclei boundaries are manually labeled by experts. These manually annotated boundaries are treated as the reference (ground truth) for the refinement evaluation. In each ROI every immunopositive and 15 randomly selected immunonegative nuclei were marked. In total, there are 600 manually marked nuclei (405 immunopositive and 195 immunonegative) in 13 images. Example of such annotation is presented in Fig. 1 (first column). Since the analyzed images of thresholding results have much more objects than the ground truth the evaluation is calculated only in the vicinity (bounding boxes) of the objects in ground truth.

3   Results

We measure the algorithm performance on microscopic images by comparison of refined boundary evaluation in relation to manually annotated ground truth. The performance is evaluated using pixel-wise sensitivity and specificity metrics.

The boundary refinement is performed on every object in results of 4 thresholding algorithms in 13 ROI. Each object undergoes boundary refinement by 5 algorithms on two monochromatic images containing portion of color information as separate dye layers of H and DAB as the result of color deconvolution. The example of threshold and refinement results are presented in Fig. 2. The comparison of mean values of sensitivity and specificity are presented in Table 1. For each image, we also compared the percentages of incorrectly labeled pixels before and after boundary refinement – Table 2.

It appears that satisfactory results were achieved only by two algorithms based on active contour – refine_AC and refine_CV. This is true for both monochromatic images of DAB and H computed with color deconvolution. In case of both methods we observe improvement in sensitivity metric with similar value of specificity in comparison to results without applied boundary refinement.

4   Discussion

In this investigation we try to prove that results of segmentation are improved with applying boundary refinement algorithm after image thresholding. Moreover, we verify how much the results could be improved. To confirm that the improvement is independent of initial thresholding technique we tested the refinement algorithms on results of 4 different methods of thresholding. Based on this comparison we observed that two of them (refine_AC and refine_CV) based on active contour improved the segmentation sensitivity in the case of all tested

Table 1. Comparison of Sensitivity and Specificity metrics before and after boundary refinement with 5 methods performed on results of 4 thresholding methods. Results presented for DAB layer and Bradley thresholding method. Improved results in bold.

		deconv_DAB		deconv_H	
		Sensitivity	Specificity	Sensitivity	Specificity
bradley	no refine	$0,901 \pm 0,054$	$0,996 \pm 0,003$	$0,901 \pm 0,054$	$0,996 \pm 0,003$
	refine_AC	**$0,929 \pm 0,045$**	$0,995 \pm 0,004$	**$0,915 \pm 0,048$**	$0,995 \pm 0,003$
	refine_CV	**$0,916 \pm 0,038$**	$0,994 \pm 0,004$	**$0,913 \pm 0,048$**	$0,995 \pm 0,003$
	refine_RG	$0,601 \pm 0,120$	$0,995 \pm 0,004$	$0,397 \pm 0,110$	$0,998 \pm 0,001$
	refine_RRG	$0,218 \pm 0,090$	$0,999 \pm 0,000$	$0,039 \pm 0,031$	$1,000 \pm 0,000$
	refine_KM	$0,831 \pm 0,087$	$0,995 \pm 0,004$	$0,849 \pm 0,073$	$0,996 \pm 0,003$
fuzzcc	no refine	$0,710 \pm 0,109$	$0,998 \pm 0,001$	$0,710 \pm 0,109$	$0,998 \pm 0,001$
	refine_AC	**$0,762 \pm 0,103$**	$0,997 \pm 0,003$	**$0,759 \pm 0,107$**	$0,997 \pm 0,003$
	refine_CV	**$0,763 \pm 0,093$**	$0,997 \pm 0,002$	**$0,757 \pm 0,105$**	$0,997 \pm 0,002$
	refine_RG	$0,610 \pm 0,079$	$0,997 \pm 0,003$	$0,455 \pm 0,161$	$0,998 \pm 0,001$
	refine_RRG	$0,236 \pm 0,114$	$1,000 \pm 0,000$	$0,047 \pm 0,032$	$1,000 \pm 0,000$
	refine_KM	$0,741 \pm 0,094$	$0,996 \pm 0,003$	$0,758 \pm 0,109$	$0,997 \pm 0,002$
mce	no refine	$0,839 \pm 0,077$	$0,997 \pm 0,002$	$0,839 \pm 0,077$	$0,997 \pm 0,002$
	refine_AC	**$0,878 \pm 0,067$**	$0,996 \pm 0,003$	**$0,868 \pm 0,070$**	$0,996 \pm 0,003$
	refine_CV	**$0,870 \pm 0,062$**	$0,995 \pm 0,004$	**$0,864 \pm 0,070$**	$0,996 \pm 0,003$
	refine_RG	$0,625 \pm 0,092$	$0,996 \pm 0,004$	$0,415 \pm 0,164$	$0,998 \pm 0,001$
	refine_RRG	$0,230 \pm 0,088$	$1,000 \pm 0,000$	$0,045 \pm 0,030$	$1,000 \pm 0,000$
	refine_KM	$0,807 \pm 0,086$	$0,995 \pm 0,004$	$0,828 \pm 0,083$	$0,997 \pm 0,003$
nick	no refine	$0,698 \pm 0,161$	$0,997 \pm 0,003$	$0,698 \pm 0,161$	$0,997 \pm 0,003$
	refine_AC	**$0,736 \pm 0,167$**	$0,996 \pm 0,004$	**$0,724 \pm 0,164$**	$0,996 \pm 0,003$
	refine_CV	**$0,726 \pm 0,176$**	$0,996 \pm 0,004$	**$0,722 \pm 0,163$**	$0,996 \pm 0,003$
	refine_RG	$0,483 \pm 0,112$	$0,995 \pm 0,005$	$0,305 \pm 0,132$	$0,998 \pm 0,002$
	refine_RRG	$0,202 \pm 0,094$	$0,999 \pm 0,000$	$0,033 \pm 0,027$	$1,000 \pm 0,000$
	refine_KM	$0,681 \pm 0,179$	$0,995 \pm 0,004$	$0,683 \pm 0,164$	$0,997 \pm 0,003$

introductory thresholds (as presented in Table 1). For each image, we also compared the percentages of incorrectly labeled pixels before and after boundary refinement. We observe consistent improvement for all thresholding methods for both active contour methods (as presented in Table 2).

It was established that two algorithms based on active contour produce improved object boundary in comparison to thresholding results without refined boundary. Indicated dependencies are true while processing both monochromatic images (DAB and H) computed with color deconvolution. Overall, we achieved better results while refining boundaries based on DAB layer.

Table 2. Comparison of erroneously labeled pixel percentage before and after boundary refinement with 5 methods performed on results of 4 thresholding methods. Improved results in bold. *[all values in percent]*

	no refine	refine_AC	refine_CV	refine_RG	refine_RRG	refine_KM
DAB						
bradley	10.93	**9.47**	**10.14**	19.57	24.43	12.40
fuzzcc	14.48	**11.30**	**11.54**	16.21	23.06	13.53
mce	11.68	**9.50**	**9.96**	17.95	24.10	12.48
nick	14.59	**11.72**	**12.43**	22.17	24.68	14.44
H						
bradley	10.93	**9.38**	**9.55**	23.58	29.06	11.18
fuzzcc	14.48	**11.72**	**11.98**	21.00	28.82	12.89
mce	11.68	**9.72**	**10.01**	22.44	28.84	11.63
nick	14.59	**12.01**	**12.25**	24.76	29.22	13.94

The refine_AC algorithm is the best performing method in the case of nuclei boundary refinement. It has best sensitivity and specificity for monochromatic images of DAB (color deconvolution) thresholded with Bradley method, 0.929 and 0.995 respectively. This result corresponds to relative improvement of 0.028 in sensitivity metric with comparative value of specificity. The mean relative improvement (for all segmentation methods) for DAB layer is 0.039 while for H layer it is 0.030. The mean improvement in erroneously labeled pixels for refine_AC algorithm is about 2.3 percent points. In comparison to related works (see Sect. 1.1) we achieved similar improvement for our type of data.

While both methods based on active contour improve the segmentation accuracy we observed that method refine_CV results in more indistinct edge – it may be a closer fit to object's boundary. Nevertheless, the refine_AC algorithm is still preferred as it gives more smooth object boundary. We believe that in this case the boundary is consistent with human perception paradigm according to which the smooth boundary is followed after initial general overview of the object.

To sum up, thresholding algorithm's results almost never perfectly fit to object's boundary, but this preliminary detection of objects followed by boundary refinement results in more accurate segmentation. Based on the results we can assume that applying boundary refining algorithm improves the segmentation results regardless of the used initial object detection method and of the chosen intensity layer (DAB or H). By applying method refine_AC we managed to improve the sensitivity. While using object boundary refinement to process high-content whole slide images to support the diagnosis this improvement might cause significant increase in correct diagnostic decisions.

Funding. We acknowledge the financial support of the Polish National Science Center grant, PRELUDIUM, 2013/11/N/ST7/02797.

References

1. Al-Kofahi, Y.: Improved automatic detection and segmentation of cell nuclei in histopathology images. IEEE Trans. Biomed. Eng. **57**(4), 841–852 (2010)
2. Bezdek, J.C.: FCM: the fuzzy c-means clustering algorithm. Comput. Geosci. **10**(2–3), 191–203 (1984)
3. Bradley, D., et al.: Adaptive thresholding using the integral image. J. Graph. Tools **12**(2), 13–21 (2007). https://doi.org/10.1080/2151237X.2007.10129236
4. Brink, A., et al.: Minimum cross-entropy threshold selection. Pattern Recogn. **29**(1), 179–188 (1996)
5. Bunyak, F., et al.: Histopathology tissue segmentation by combining fuzzy clustering with multiphase vector level sets. In: Advances in Experimental Medicine and Biology, pp. 413–424. Springer, New York (2011)
6. Daniel: Implementation of recursive region growing algorithm (...) (2011). https://www.mathworks.com/matlabcentral/fileexchange/32532. Accessed 18 June 2018
7. Fonseca, P.: Implementation of k-means image segmentation based on histogram (2011). https://www.mathworks.com/matlabcentral/fileexchange/30740. Accessed 18 June 2018
8. Irshad, H.: Methods for nuclei detection, segmentation, and classification in digital histopathology: a review–current status and future potential. IEEE Rev. Biomed. Eng. **7**, 97–114 (2014)
9. Khurshid, K., et al.: Comparison of niblack inspired binarization methods for ancient documents. In: Document Recognition and Retrieval XVI (2009)
10. Korzynska et al., A.: The METINUS plus method for nuclei quantification in tissue microarrays of breast cancer and axillary node tissue section. Biomed. Signal Process. **32**, 1–9 (2017)
11. Kroon, D.J.: Implementation of segmentation by growing a region from seed point using intensity mean measure (2008). https://www.mathworks.com/matlabcentral/fileexchange/19084. Accessed 18 June 2018
12. Markiewicz, T., et al.: MIAP – web-based platform for the computer analysis of microscopic images to support the pathological diagnosis. Biocybern Biomed. Eng. **36**(4), 597–609 (2016)
13. Pang, J.: Implementation of the localized active contour method using level set method (2014). https://www.mathworks.com/matlabcentral/fileexchange/44906. Accessed 18 June 2018
14. Roszkowiak, L., et al.: Nuclei detection methods in immunohistochemically stained tissue sections of breast cancer. (Submitted to Computer Methods and Programs in Biomedicine on 19 June 2018)
15. Snoj, N., et al.: Molecular Biology of Breast Cancer, pp. 341–349. Academic Press (2010). https://www.scopus.com/inward/record.uri?eid=2-s2.0-84884578034&doi=10.1016%2fB978-0-12-374418-0.00026-8&partnerID=40&md5=75f68bd64c6de8f9dde1630b973b017d
16. Wu, Y.: Implementation of active contours without edges by Chan and Vese (2009). https://www.mathworks.com/matlabcentral/fileexchange/23445. Accessed 18 June 2018

Football Players Pose Estimation

Michał Sypetkowski[1,2], Grzegorz Kurzejamski[2], and Grzegorz Sarwas[2,3(✉)]

[1] Institute of Computer Science, Warsaw University of Technology, Warsaw, Poland
m.sypetkowski@stud.elka.pw.edu.pl
[2] Sport Algorithmics and Gaming, Warsaw, Poland
{m.sypetkowski,g.kurzejamski,g.sarwas}@sagsport.com
[3] Institute of Control and Industrial Electronics,
Warsaw University of Technology, Warsaw, Poland
sarwasg@ee.pw.edu.pl

Abstract. The paper presents analysis of algorithms for football players pose estimation based on a custom, real scenario data. Listed approaches have been examined on high resolution videos or photos taken from multiple cameras during football match or training. Wide views has been considered, producing extremely low resolution players' visuals. Tested algorithms consisted of retrained deep convolutional models, optionally retrained for different generalization scenarios. Multiple scenarios for pose estimation compliance have been proposed as well as options for further model augmentations and retraining.

1   Introduction

Visual analysis of football matches and training sessions is a demanding tasks, consisting of multiple aspects as proper video acquisition, tracking in a multi-view system with occlusions, 3D calibration and human behavior analysis. The latter can be split in various conceptual and algorithmic problems, one of each is player's pose estimation. Human pose helps football analysts to validate players' mobility during match and ability to properly perform various game interceptions. In particular analysts checks how often player uses non-dominant leg during ball repossession. Accurate pose estimation is also a key step for higher level tasks as analysis of visibility of action for each player.

Using multiple 2D pose estimations on images taken from multiple cameras, it is possible to reconstruct 3D skeleton using off-the-shelf algorithms. For such task, there must be at least 2 calibrated cameras. In general: the more cameras, the more errors in 2D pose estimation are allowed for precise reconstruction.

Visual tracking systems installed in football academies uses wide view cameras, spanning on whole pitch or near half. Depending on installation site cameras could be positioned near ground, producing substantial occlusions, or on a high pylons giving non-standard human view from above. Moreover, wide view cameras imposes very low-quality human visuals even for top tier recording hardware. We did not find any literature nor the databases with annotated human pose for the high view and low resolution scenario, what imposed the research problem.

© Springer Nature Switzerland AG 2019
M. Choraś and R. S. Choraś (Eds.): IP&C 2018, AISC 892, pp. 63–70, 2019.
https://doi.org/10.1007/978-3-030-03658-4_8

All successful pose estimation approaches concern high or medium resolution images. The literature presents two generalized approaches in that case. The first one is called bottom-up and the second is top-down. We tested multiple known state-of-the-art algorithms for pose estimation with our custom test images. The images have been acquired form real system with four high-view and high-class wide-view cameras. In next subsections we present analysis of related work in different pose estimation approaches.

1.1   2D Multi-person Bottom-Up Approaches

Bottom-up approach predicts all keypoints, which are considered as skeleton model parts in a single scene. Those are further assembled into full skeleton by assigning the parts to appropriate place in the model. In [3] the multiple-stage fully convolutional networks for estimating Part Confidence Map (heat map) and PAF (Part Affinity field - 2D vector field) have been considered. This solution uses multi-stage convolutional network that generates heat map and 2D vector field for each body part (e.g. right elbow, left wrist, neck). The affinity graph is build using 2D vector field part. Based on it, the 2D skeleton with a particular heuristic graph relaxation technique proposed in the article can be constructed. The approach presented in [11] achieved the best result in COCO 2016 Keypoint Detection Task, being valid proposition for solving our problem. Along with work of Simon [16], this approach has publicly available implementation called Open-Pose [8]. Highest score on MPII multi-person pose dataset [1] got an approach presented in work of Newell et al. [13]. Authors trained a network to simultaneously output detections and group assignments. Output of their neural network consist of detection heatmaps with respective associative embeddings. Grouping body parts is performed by an algorithm based on thresholding the parts embeddings distances. This approach differs from other bottom-up approaches in that there is no separation of a detection and a grouping stage. An entire prediction is done at once by a single-stage, generic network. The network is based on a stacked hourglass architecture [14].

1.2   2D Multi-person Top-Down Approaches and Single Person Pose Estimation

Top-down approaches localize and crop all persons from an image at first, then solve the single person pose estimation problem (which becomes the main difficulty). Modern single person pose estimation techniques incorporate priors about the a structure of human bodies. Best results in COCO 2017 Keypoint Detection Task [11] were achieved by Cascaded Pyramid Network [4]. This work focuses on the "hard" keypoints (i.e. occluded, invisible and with non-trivial background). It is achieved by explicitly selecting the hard keypoints and backpropagating the gradients from the selected keypoints only.

Approach called Mask R-CNN [7], extends Faster R-CNN [15] by adding a branch for predicting an object mask in parallel with bounding box recognition. Using this simple modification the Mask R-CNN can be be applied to detect keypoints. This approach achieves high results in all COCO 2017 challenges (i.e. object detection, object segmentation, keypoint detection).

Paper by Simon et al. [16] presents precise hand 2D keypoint detector. It introduces a semi-supervised training algorithm called Multiview Bootstrapping. Initially, the algorithm needs a set of annotated examples. The model is trained using only these examples at the beginning. Then, the model detects keypoints on unannotated examples with multiple camera views. Each multi-view example is then robustly 3D triangulated, and reprojected creating additional training set.

Stacked hourglass [14] achieves state-of-the-art result on MPII [1]. It presents a CNN architecture for bottom-up and top-down inference with residual blocks. Approach introduced by Ke et al. [9] aims to improve stacked hourglass [14]. It achieves currently best score on MPII single person pose dataset.

Adversarial PoseNet [5] paper presents an interesting approach that trains a GAN, with multi-task pose generator and two discriminator networks. It achieves state-of-the-art results on MPII [1] single person pose estimation dataset. The model consists of the generator network, the pose discriminator network and the confidence discriminator. Half of them represent keypoint locations and the other half occlusion predictions. The generator architecture is based on stacked hourglass architecture [14]. Despite significant differences in model and training method, it's result is not that significantly better on MPII than original hourglass approach (+1.0 PCKh).

2   Pose Estimation for Football

As a part of a complex system, automation of the analysts' work often considers various aspects of player's pose. For instance, torso's direction defines mobility levels for different directions and visibility scores. During interaction with the ball it is important which leg is used the most. In ideal situation the head direction could be of significant value to the analysts. These aspects needs kind of positional data presented in 3D world. Given the multiple camera system, 3D pose can be achieved by triangulating the skeleton keypoints. For that to happen, the skeletons should present enough accuracy not to miss classify legs for instance. Different body parts has different visual saliency and thus are differently assessed by algorithms. Head direction for instance is not achievable by any means with use of tested algorithms, as facial features hardly gives any response in the image's signal.

For our tests we gathered data using 4 cameras placed at the field corners. Because of resolution limits, in practice we can assume that for a given player only 2 cameras are close enough to produce usable visuals. The cameras are production class CCTV devices with 4 K resolution and high compression bandwidth. Even though the crop factor around single player magnifies compression and optics artifacts, which renders high frequency data unusable. Low quality and viewing angle creates uncommon characteristics of the images. Comparing this scenario with to standard pose estimation datasets like COCO [11] and MPII [1] we can list main problems:

- Human based annotations are much more difficult and time consuming for our images. Some images have practically indistinguishable joint locations, even with much human time and effort spent
- Border areas of the pitch generates almost top down views, where the human parts are mostly occluded by upper body
- Images are blurred with non-deterministic distribution, which makes generic upscaling algorithms useless
- All players wear single-color clothes, which makes it harder to distinguish limbs (especially hands) from the body

3   Experiments

We tested current state-of-the art approaches, and measured their performance on our data. To check their robustness tested 5 implementations on 300 examples of our images containing one player. On top of that we performed two experiments with retraining existing models and augmenting our image database with GANs.

3.1   Base Models

Implementations vary in skeleton structure used as a reference. We have no annotations for our images, as the problem presented in the paper is too complex to incorporate single accuracy metric. We've taken into account few the easiest football aspects for automation. We measured precision with human based decision, whether the answer is one of 4 classes: correct, only correct legs pose estimation, wrong pose, N/A. The human-based bias has been lowered by cross-checkup with industry football analyst but still may produce significant variance, opposed to keypoint-based difference metrics.

The results are shown in Table 1. Example detections of selected implementations are shown in Fig. 1. OpenPose [8], achieved low precision (only 58/300 correct whole poses and 106/300 correct legs pose estimations). This approach is designed for multi-person pose estimation, therefore it cannot assume that there is exactly one whole person on the image and that it is centered. Best results were achieved by CPN [4], which outperforms stacked hourglass [14] by 10% on whole pose estimation, and by 7% on legs pose estimation.

3.2   Improving the Results

We used data augmentation method so that the distribution of training examples of a model would be more similar to ours. We used various blur algorithms and noise applied randomly with random amplitudes. Example of such augmentation is shown in Fig. 2. We retrained CPN [4] (with initial weights restored from the

Table 1. Selected human pose estimation implementation results (original and our experiments)

Approach	Implementation/ experiment	Training set	Language, library	Corr. pose	Corr. legs	N /A
PAF [3]	OpenPose[a]	COCO [11]	C++, Caffe	58	106	29
Stacked hourglass [14]	original implementation[b], 8-stack model	MPII [1]	Lua, Torch	142	203	-
	alternative implementation[c], hg_refined_200, 4-stack model,	MPII [1]	Python, Tensor-flow	29	90	
	alternative implementation - not official[d], 8-stack model,	MPII [1]	Python, Pytorch	135	186	
CPN [4]	Original implementation[e], COCO.res50.256x192, snapshot_350.ckpt	COCO [11]	Python, Tensor-flow	171	**224**	-
	SRGAN for upscaling			90	167	
	blurred images, 50 more epochs, lr 1.6e-5 (from COCO.res50.256x192)			155	206	
	COCO.res101.384x288, snapshot_350.ckpt			158	223	

[a] https://github.com/CMU-Perceptual-Computing-Lab/openpose
[b] https://github.com/umich-vl/pose-hg-demo
[c] https://github.com/wbenbihi/hourglasstensorlfow
[d] https://github.com/bearpaw/pytorch-pose
[e] https://github.com/chenyilun95/tf-cpn

original checkpoint), using such data augmentation. In the end we didn't achieve better results (see Table 1).

It seems that the model is already trained so that it can handle even low-resolution images. We hypothesize, that good generalization for low-resolution images, may be caused by the network structure that produces intermediate feature maps (each used to produce intermediately supervised heatmaps) in various resolutions. In the end it combines and upscales heatmaps achieved on all levels.

Another experiment we performed is using SRGAN [10] for upscaling our test images. The upscaling caused only unwanted artifacts and even decreased pose estimation precision (by 27% on whole pose, and by 19% on legs). An example upscaling effect is shown Fig. 3. Given the results one can wonder if the images are too small for SRGAN to work properly or the training data doesn't represent our scenario.

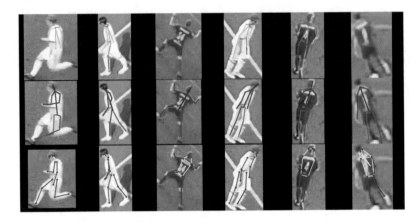

Fig. 1. Example detections for selected implementations (row 1—OpenPose, row 2—Stacked hourglass (original implementation), row 3—CPN)

Fig. 2. Example of data augmentation (on COCO dataset) that we used in the experiment. Original image is first from the left.

Fig. 3. Left—original fragment, right—fragment 4x upscaled with SRGAN [10]

4   Conclusions and Future Work

Modern pose estimation approaches are already robust to blurring and low-resolution in general. Significantly improving their performance with simple methods, like heuristic data augmentation or upscaling the images with generic upscaling algorithms may be extremely hard with limited training data. Therefore, in this section we present case-specific ideas for achieving more precise pose estimation on our data.

4.1   Training Additional Layers and Fine Tuning

A straightforward solution for improving the results may be manual annotation of some examples (e.g. a few thousands) for training existing state-of-the-art models. Such training examples could be used for training layers at the end of the model, and for fine tuning with a small learning rate.

4.2   Avoiding Manual Annotation

Human annotations on low resolution images not only require immense amount of work, but also may be hard to be done precisely in our case. Super-FAN [2] learns to detect landmarks on small images which human is incapable to annotate. We suspect, that the blur characteristics (its distribution) may be learned by a specialized neural network. We don't have high-resolution equivalents for our images, so our choices are:

1. using existing high-resolution datasets (downscaling images for training)
2. synthetic dataset
3. unsupervised learning (eventually semi-supervised e.g. [16])

3D Hand pose estimation approach [12] focuses on enhancing a synthetic dataset to make their distribution more like the distribution of the real images. It uses CycleGAN [17] with an additional geometric consistency loss. The paper shows, that training with generated images significantly outperform standard augmentation techniques. Similar approach may be applied to pose estimation. In our case, use of silhouettes (geometric consistency loss) may turn out to be less efficient, because our real images are blurry and in low resolution. Because of the blur, we may experiment with partially transparent silhouettes (grayscale). With low-resolution images, synthetic training set seems to be even more appropriate in our case. The human body details are not visible on our images, therefore GANs are easier to learn their distribution.

4.3   Upscaling

Modern generic upscaling deep learning methods are focused on minimizing the mean squared reconstruction error (MSE) [6]. SRGAN [10] is capable of inferring photo-realistic natural images for 4x upscaling factors. Such generic upscaling algorithms like these won't improve results on our dataset, because of too low resolution and characteristic distortions. To address problem of very low resolution, an approach worth trying is adjusting Super-FAN [2], so that it detects pose keypoints instead of facial landmarks.

Acknowledgements. This work was co-financed by the European Union within the European Regional Development Fund.

References

1. Andriluka, M., Pishchulin, L., Gehler, P., Schiele, B.: 2D human pose estimation: new benchmark and state of the art analysis. In: Proceedings of IEEE Conference Computer Vision and Pattern Recognition (CVPR), pp. 3686–3693, June 2014
2. Bulat, A., Tzimiropoulos, G.: Super-FAN: integrated facial landmark localization and super-resolution of real-world low resolution faces in arbitrary poses with GANs. arXiv preprint arXiv:1712.02765 (2017)
3. Cao, Z., Simon, T., Wei, S.E., Sheikh, Y.: Realtime multi-person 2D pose estimation using part affinity fields. In: Proceedings of IEEE Conference Computer Vision and Pattern Recognition (CVPR), pp. 1302–1310, July 2017
4. Chen, Y., Wang, Z., Peng, Y., Zhang, Z., Yu, G., Sun, J.: Cascaded pyramid network for multi-person pose estimation. arXiv preprint arXiv:1711.07319 (2017)
5. Chen, Y., Shen, C., Wei, X.S., Liu, L., Yang, J.: Adversarial PoseNet: a structure-aware convolutional network for human pose estimation. CoRR, abs/1705.00389 2 (2017)
6. Dong, C., Loy, C.C., He, K., Tang, X.: Image super-resolution using deep convolutional networks. IEEE Trans. Pattern Anal. Mach. Intell. **38**(2), 295–307 (2016)
7. He, K., Gkioxari, G., Dollár, P., Girshick, R.: Mask r-CNN. In: Proceedings of IEEE International Conference Computer Vision (ICCV), pp. 2980–2988, October 2017
8. Hidalgo, G., Cao, Z., Simon, T., Wei, S.E., Joo, H., Sheikh, Y.: Openpose, June 2017. https://github.com/CMU-Perceptual-Computing-Lab/openpose
9. Ke, L., Chang, M.C., Qi, H., Lyu, S.: Multi-scale structure-aware network for human pose estimation. arXiv preprint arXiv:1803.09894 (2018)
10. Ledig, C., Theis, L., Huszár, F., Caballero, J., Cunningham, A., Acosta, A., Aitken, A., Tejani, A., Totz, J., Wang, Z., Shi, W.: Photo-realistic single image super-resolution using a generative adversarial network. In: Proceedings of IEEE Conference Computer Vision and Pattern Recognition (CVPR), pp. 105–114, July 2017
11. Lin, T.Y., Maire, M., Belongie, S., Hays, J., Perona, P., Ramanan, D., Dollár, P., Zitnick, C.L.: Microsoft COCO: common objects in context. In: European conference on computer vision, pp. 740–755. Springer (2014)
12. Mueller, F., Bernard, F., Sotnychenko, O., Mehta, D., Sridhar, S., Casas, D., Theobalt, C.: Ganerated hands for real-time 3D hand tracking from monocular RGB. CoRR abs/1712.01057 (2017). http://arxiv.org/abs/1712.01057
13. Newell, A., Huang, Z., Deng, J.: Associative embedding: end-to-end learning for joint detection and grouping. In: Advances in Neural Information Processing Systems, pp. 2277–2287 (2017)
14. Newell, A., Yang, K., Deng, J.: Stacked hourglass networks for human pose estimation. In: European Conference on Computer Vision, pp. 483–499. Springer (2016)
15. Ren, S., He, K., Girshick, R., Sun, J.: Faster r-CNN: towards real-time object detection with region proposal networks. IEEE Trans. Pattern Anal. Mach. Intell. **39**(6), 1137–1149 (2017)
16. Simon, T., Joo, H., Matthews, I., Sheikh, Y.: Hand keypoint detection in single images using multiview bootstrapping. In: Proceedings of IEEE Conference Computer Vision and Pattern Recognition (CVPR), pp. 4645–4653, July 2017
17. Zhu, J.Y., Park, T., Isola, P., Efros, A.A.: Unpaired image-to-image translation using cycle-consistent adversarial networks. arXiv preprint (2017)

Text Entry by Rotary Head Movements

Adam Nowosielski[(✉)]

West Pomeranian University of Technology, Szczecin, Faculty of Computer Science
and Information Technology, Żołnierska 52, 71-210 Szczecin, Poland
anowosielski@wi.zut.edu.pl

Abstract. Head movements are one of several possibilities to operate electronic equipments or computers by physically challenged people.
A camera mouse where users control the on-screen pointer through the
head movements is now a standard which has many implementations.
The process of typing, however, in such interfaces still remains a challenge. In this paper, a gesture-based technique is adopted for touchless typing with head movements. To type a character a rotary head
movement is employed. The changes in the direction pattern are used
for accessing consecutive characters. Quick access to individual letters,
elimination of the pressing (or its simulation) and no need for precision
during typing are the main advantages of the proposed approach.

Keywords: Human-computer interaction · Touchless typing
Typing with head movements · Virtual keyboard

1 Introduction

The QWERTY key arrangement has been designed for ten fingers typing and is
not well suited for different modes of accessing keys. Even for typing with the
fingers the QWERTY keyboard is argued to had been invented to be slow to
prevent jams on mechanical typewriters. Many alternatives have appeared but
none has gained a widespread popularity and recognition. This applies both to
physical full-size keyboards (like Dvorak or Colemak) and to virtual on-screen
keyboards (like Fitaly or Opti) of the mobile devices with touch screens. Users
are reluctant to learn new key arrangements. For that reason new approaches
that accelerate typing on the QWERTY keyboard have begun to appear with
the swipe technique being the most prominent example here.

Thanks to the new typing techniques and the dictionary support, despite
its inherent disadvantages, the QWERTY keyboard will probably remain a non-
threatened standard for the general public. For some people, however, who suffer
for different form of physical disability the QWERTY keyboard might not be
the best suited solution. When considering the head operated interface many
factors are against the discussed layout.

The head operated interfaces are an option for people with physical hand-
icap. Other alternatives include [1]: hand gesture recognition, brain computer

© Springer Nature Switzerland AG 2019
M. Choraś and R. S. Choraś (Eds.): IP&C 2018, AISC 892, pp. 71–78, 2019.
https://doi.org/10.1007/978-3-030-03658-4_9

interfaces, eye tracking, speech recognition, lip movement analysis (silent speech recognition) and others. The choice of the appropriate solution depends predominantly on the form of disability. For the user it should improve the level of independence in everyday life, allow participation in social activities, offer means for operation in the electronic world.

In the paper, a gesture-based technique is adopted for touchless typing with head movements. Each character is entered by continuous rotary head movement (a gesture) which is natural and easy to make by an user. Moreover, it does not require a precision from the person performing a gesture since there is no pointing procedure involved. The approach does not require a simulation of the clicking event (key pressing), neither.

The paper is structured as follows. In Sect. 2 the process of typing in head operated interfaces is referenced. Then in Sect. 3 the concept of rotary movement typing is introduced. Text entry by rotary head movements is proposed in Sect. 4. The interface is evaluated in Sect. 5. Final conclusions and a summary are provided in Sect. 6.

2   Typing in Head Operated Interface

In the head operated interface, typing can be available for the user through the on-screen virtual QWERTY keyboard operated by the camera mouse technique (e.g. [2,3]). It works on the following routine. The face of the user is continuously tracked and the changes are mapped to the position of the pointer in the Graphical User Interface (GUI). To type, a user moves the pointer above the intended letter on the virtual keyboard and perform the pressing action. The pressing requires implementation of additional mechanisms and most popular solutions include: eye blinks [3–5], mouth shape changes (opening, closing or stretching) [2,6–8], brows movements [7] or cheeks twitch. The hover to selects technique, which requires to hold the pointer for a moment over a letter or other object being pointed, is also practically used. Its main disadvantage is prolongation of the text entry process. The introduction of swipe techniques in the camera mouse approach can partially solve the problem. However, some actions are still requested for distinguishing the beginning and ending of the swipe gesture.

Another possibility of typing with the head movement using the on-screen QWERTY keyboard is the traverse procedure. In the most basic form, one key is active (highlighted) and others are accessed in a sequence according to the direction of the head movement. In the traverse procedure, when subsequent characters are distant from each other, many steps (i.e. head movements) might be required to reach the next letter. When the user finally reaches the requested character a pressing action is needed. It is performed similarly as described in the above paragraph (i.e. using eye blinks, opening the mouth, cheeks twitch, etc.).

There are some examples in the literature that modify the above scheme. In [9] only directional movements are used for typing. The traverse procedure uses only left and right directional movements and consists of two steps: selecting a column and then a row. Downward movement is reserved for the pressing event and upward movement for the cancellation. In the Assistive Context-Aware Toolkit (ACAT) [10] keys on the QWERTY keyboard are grouped and accessed in a cyclic mode. The grouping results in faster reaching of the target character. The cyclic routine can also be performed automatically. This provides the solution which can be operated with only two state input signal (e.g. eye blink).

As can be observed from the above brief description typing through head movements using the QWERTY keyboard in touchless mode is not convenient. The process requires substantial precision with the pointing procedure or is time-consuming when cycling through adjacent characters in the traverse routine.

3   Rotary Movement Typing

Despite the fact that many alternatives to the QWERTY key arrangement have failed for the general public, they may prove their value in a very specific head operated interface. A wide overview of the existing keyboard layouts is presented in [11]. In our earlier studies we substituted the QWERTY layout with a single row alphabetically arranged characters for touchless head typing [14]. This approach required only directional head movements for operation. Later, we proposed the 3-Steps Keyboard, a hierarchical variant of a single row keyboard, where each letter is accessed with a sequence of three directional head movements. What we had discovered during experiments was the observation that users can perform, quite easily, movements in other than main directions (left, right, up and down) but those movements lack precision. They, however, can be implemented in the interface and offer additional value. In this paper, we are focusing our attention on the rotary head movements which allow different manner of accessing characters. Our goal is to obtain the head typing mechanism similar to that found in the Quikwriting [12] or the 8pen [13] (see Fig. 1) both designed for the touch screens and mobile devices.

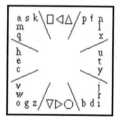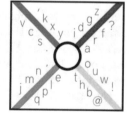

Fig. 1. Two variants of the Quikwriting [12] keyboard layouts (left and middle) and the 8pen [13] layout (right)

In the Quikwriting there are 9 zones, arranged in 3×3 grid. The central zone is neutral. It is a resting zone from which interaction begins. The user drags the stylus out to one of the eight outer zones, then optionally to neighbouring outer zones, and finishes in the central zone [12]. Moving to the outer zone and return corresponds to the character in the middle of the given zone. Optional movement to the neighbouring zones denotes consecutive characters from the starting zone. This approach allows its user to write entire words in a single continuous gesture [12].

The concept of the 8pen solution is much similar to the Quikwriting. The division of the plane into sectors is the main difference. There are 5 regions in the 8pen: 1 central and 4 directional. The central part is the resting zone equivalent to the starting/ending point. Movement to the directional sector limits potential characters to those placed in that region. A choice of the particular letter is based on further circular movement through remaining sectors. The number of sectors crossed corresponds to the position of the letter on the border of the entry sector. The process finishes in the central region. It is symbolically depicted in Fig. 2. The red dotted lines denote the trail of the movements. There are three examples representing the process of entering three letters: 'a', 'd', and 'm'. In each case the procedure begins in the middle. The 'a' character is placed in the upper part of the right zone. This means that after moving to the right sector, the movement must continue upward. Since the 'a' letter is the first from the center, only one sector must be crossed. To type 'd', which is the second letter on the right hand side in the upper zone, two sectors ought to be crossed. Similarly with the 'm' character which requires crossing of the three sectors.

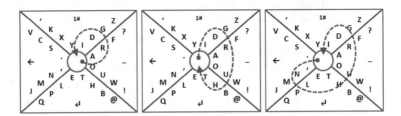

Fig. 2. Examples of entering letters on the 8pen [13] keyboard

As can be observed from examples presented in Fig. 2 each sector have additional character placed in the middle near the border. We added them to gain the quick access to space bar (right), backspace (left), enter (down), and additional characters not included in the standard layout (up). Those keys are accessed with directional movement from the central zone to the appropriate sector and return to the central region.

4   Text Entry by Rotary Head Movements

Text entry methods introduced by the Quikwriting and the 8pen are designed for touch screens. They can be adopted, however, to the touchless mode. Beside the obvious disadvantage: the necessity to learn new key arrangements, these techniques offer the following advantages:

- each character is entered with a continuous movement;
- typing a word is possible by a single continuous gesture;

- precision is not required (no need to aim at a small key on the keyboard);
- there is a neutral area (central zone) which can naturally be used for resting;
- access to individual letters is possible without pressing;
- frequent characters (for the English language) are accessed rapidly;
- frequent bigrams (for the English language) can by entered by a flow movement.

The rotary head operated interface proposed in this paper is based on the 8pen solution. It seems that this approach requires less precision than the Quikwriting. Starting from the central sector, eight directions should be recognized in the Quikwriting while in the 8pen there are only four.

The proposed interface is presented in Fig. 3 on the left. There is a dot (pointer) in the middle representing the position of the users's head. From the user's perspective a control is performed by moving the dot over the appropriate sectors with the movement of the head. The exemplary motion path during 'computer' word entry is presented on the right hand side in Fig. 3.

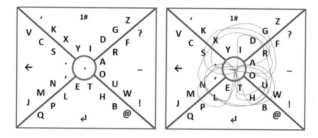

Fig. 3. The interface (left) and exemplary motion path during 'computer' word entry with head movements (right)

The control routine is based on recognition of the user actions performed with the head. The task include face detection, followed by a tracking procedure. The problem of human face detection in static images is quite well researched [16] and there are ready to use solutions available. They offer very good accuracy of the estimation of the head position [16]. The procedure of face detection and tracking in the presented solution is a classical one. We have used it before in the 3-Steps Keyboard proposed in [15]. It consists of well known Viola and Jones approach [17] followed by the Kanade-Lucas-Tomasi (KLT) feature-tracking procedure [18]. The procedure uses distinctive point selection algorithm developed by Shi and Tomasi [19]. The cited approach occurred to be unstable in the context of rotary head movements. To solve the problem we reduced the tracking area from the whole face to the small region containing the nose.

5 Evaluation

To validate the proposed approach the user-based experiments have been performed. The experiments have been conducted by a group of 8 persons (4 students and 4 employees of our university). Each participant has the task of writing a single word ('computer') in the forced error correction routine [20] which obliges participant to correct each error. During experiments the time of typing was measured and CPM (chars per minute) calculated. It must be noticed, that the forced error correction condition ensures the correctness of the transcribed text but extends the process of typing. When user performs many errors the resultant CPM measure decreases. This is particularly important for the interface which is handled in a completely non-standard way (rotary movements). Here, none of the participants have previous contact with such approach. To simulate the process of learning new key arrangements and handling the interface the evaluations consisted of 10 sessions preceded by a single, reconnaissance trial carried out without time measurements. The results are presented in a graphical form in Fig. 4. The solid lines correspond to individual participants while the bar chart represents the average CPM value.

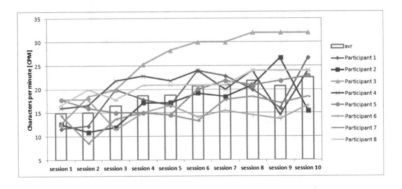

Fig. 4. The results of conducted experiments

The obtained results are quite optimistic. They show that after a number of repetitions users get acquainted with the interface and could type known word with a 22.5 CPM average. The best participant (no 3) achieved 32 CPM. Interestingly, the 15 CPM average for the first session, when the interface was a novelty for the users, is also of high value. For most users, the progress is clearly visible. The exception is participant no 6, who recorded the fastest times at the beginning (2 first session). The number of errors committed during typing is presented in Fig. 5 in a stacked bar chart. It can be observed that more errors have been committed during the initial sessions.

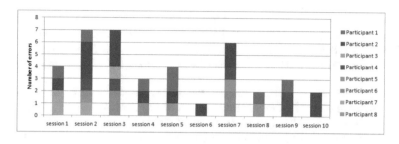

Fig. 5. The level of errors committed during experiments

6   Conclusions

In the paper, the problem of text entry using rotary head movements have been addressed and the appropriate interface proposed. The solution is based on the known concept of rotary movement used so far in touch screen interfaces. The main contribution of the paper is the adaptation of this concept to head operated interface and its evaluation. The solution proposed, proved to operate stable and the results obtained are promising. The access to individual letters is quick through the rotary head movements and there is no need for implementation of confirmation/pressing event. The movements users perform do not have to be precise.

References

1. Kumar, D.K., Arjunan, S.P.: Human-Computer Interface Technologies for the Motor Impaired, 214 p. CRC Press (2015)
2. Tu, J., Tao, H., Huang, T.: Face as mouse through visual face tracking. Comput. Vis. Image Underst. **108**(1–2), 35–40 (2007)
3. Nabati, M., Behrad, A.: 3D Head pose estimation and camera mouse implementation using a monocular video camera. Signal Image Video Process. **9**(1), 39–44 (2015)
4. Santis, A., Iacoviello, D.: Robust real time eye tracking for computer interface for disabled people. Comput. Methods Programs Biomed. **96**(1), 1–11 (2009)
5. Varona, J., Manresa-Yee, C., Perales, F.J.: Hands-free vision-based interface for computer accessibility. J. Netw. Comput. Appl. **31**(4), 357–374 (2008)
6. Bian, Z.-P., Hou, J., Chau, L.-P., Magnenat-Thalmann, N.: Facial position and expression-based human-computer interface for persons with Tetraplegia. IEEE J. Biomed. Health Inform. **20**(3), 915–924 (2016)
7. Gizatdinova, Y., Spakov, O., Surakka, V.: Face Typing: Vision-Based Perceptual Interface for Hands-Free Text Entry with a Scrollable Virtual Keyboard. IEEE Workshop on Applications of Computer Vision, Breckenridge, CO, USA, pp. 81–87 (2012)
8. Shin, Y., Ju, J.S., Kim, E.Y.: Welfare interface implementation using multiple facial features tracking for the disabled people. Pattern Recognit. Lett. **29**(2008), 1784–1796 (2008)

9. Nowosielski, A., Chodyła, L.: Touchless input interface for disabled. In: Burduk, R., et al. (eds.) Proceedings of the 8th International Conference on Computer Recognition Systems CORES 2013, AISC, vol. 226, pp. 701–709 (2013)
10. Intel Corporation: Assistive Context-Aware Toolkit (ACAT) (2018). https://01.org/acat
11. Rick, J.: Performance optimizations of virtual keyboards for stroke-based text entry on a touch-based tabletop. In: Proceedings of the 23rd Annual ACM Symposium on User Interface Software and Technology, pp. 77–86. ACM (2010)
12. Perlin, K.: Quikwriting: Continuous stylus-based text entry. In: Proceedings of the 11th Annual ACM Symposium on User Interface Software and Technology, pp. 215–216. ACM (1998)
13. 8pen (2018). http://www.8pen.com
14. Nowosielski, A.: Minimal interaction touchless text input with head movements and stereo vision. In: Chmielewski, L.J., Datta, A., Kozera, R., Wojciechowski, K. (eds.) Computer Vision and Graphics. LNCS, vol. 9972, pp. 233–243 (2016)
15. Nowosielski, A.: 3-steps keyboard: reduced interaction interface for touchless typing with head movements. In: Kurzynski, M., Wozniak, M., Burduk, R. (eds.) Proceedings of the 10th International Conference on Computer Recognition Systems CORES 2017. AISC, vol. 578, pp. 229–237 (2018)
16. Forczmański, P.: Performance evaluation of selected thermal imaging-based human face detectors. In: Kurzynski, M., Wozniak, M., Burduk. R. (eds.) Proceedings of the 10th International Conference on Computer Recognition Systems CORES 2017. AISC, vol. 578, pp. 170–181 (2018)
17. Viola, P., Jones, M.: Robust real-time face detection. Int. J. Comput. Vis. **57**(2), 137–154 (2004)
18. Lucas, B.D., Kanade, T.: An iterative image registration technique with an application to stereo vision. In: Proceedings of the 7th International Joint Conference on Artificial Intelligence, vol. 2, pp. 674–679 (1981)
19. Shi, J., Tomasi, C.: Good features to track. In: Proceedings CVPR 1994, pp. 593–600 (1994)
20. Arif, A.S., Stuerzlinger, W.: Analysis of text entry performance metrics. In: 2009 IEEE Toronto International Conference Science and Technology for Humanity (TIC-STH), pp. 100–105 (2009)

Modelling of Objects Behaviour for Their Re-identification in Multi-camera Surveillance System Employing Particle Filters and Flow Graphs

Karol Lisowski[✉] and Andrzej Czyżewski

Gdansk University of Technology, G. Narutowicza 11/12, 80-233 Gdansk, Poland
{lisowski,andcz}@sound.eti.pg.gda.pl

Abstract. An extension of the re-identification method of modeling objects behavior in muti-camera surveillance systems, related to adding a particle filter to the decision-making algorithm is covered by the paper. A variety of tracking methods related to a single FOV (Field of Vision) are known, proven to be quite different for inter-camera tracking, especially in case of non-overlapping FOVs. The re-identification methods refer to the determination of the probability of a particular object's identity recognized by a pair of cameras. An evaluation of the proposed modification of the re-identification method is presented in the paper, which is concluded with an analysis of some comparison results brought by the methods implemented with and without a particle filter employment.

Keywords: Re-identification · Multi-camera · Surveillance

1 Introduction

Automated processing of video data allows for detection and for tracking of objects as well as for interpreting of various events caused by moving objects in a given camera's FOV (Field of View). Multi-camera tracking is applied in order to support the operator in analyzing video images from many cameras which can be, in turn, distributed over a certain area. The key method of multi-camera tracking is called re-identification. The paper presents the modification of re-identification method previously developed by authors. The main challenge in evaluation of re-identification methods is related to the dependency on the results obtained from the algorithms for the analysis of video data related to a single camera image. Therefore, in order to evaluate the re-identification properly, a possible incorrectness of these algorithms has to be taken into account. In Sect. 2 works related to the topic of the paper are mentioned. Sect. 3 contains a description of the modification introduced to the re-identification method. Subsequent Sect. 4 is devoted to presentation of experiments and to comparison of results. The paper is concluded with some general remarks contained in Sect. 5.

© Springer Nature Switzerland AG 2019
M. Choraś and R. S. Choraś (Eds.): IP&C 2018, AISC 892, pp. 79–86, 2019.
https://doi.org/10.1007/978-3-030-03658-4_10

2   Related Works

In order to obtain input data for the re-identification method, the video data from each single camera have to be processed. This analysis consists of the following steps: background subtraction, object detection and tracking, event detection. Background subtraction methods are related to distinguishing between stable region of video image and regions (called also as blobs) which differ from the background. Various method and approaches (i.e. utilization of the Gaussian Mixture Model [17] or the codebook algorithm [12]) are described in literature [3,18].

Having the blobs found within a video image, moving objects can be detected and tracked. In case of methods described in this paper, the modified Kalman filters were used, which exploit our previous experience. As a result of this step of the video analysis, trajectories of a moving object can be obtained. Moreover, those methods are widely reflected in the literature [4].

The event detection algorithms are mostly based on trajectories of tracked objects. In order to detect an event in video data, some rules describing this event are necessary. Types of events can refer to the crossing barrier, entering certain area or moving towards a given direction. In the literature, methods for detecting various types of events are provided [8].

Multi-camera video analysis utilizes information related to time and location related to an appearance or a disappearance of the tracked object. Many approaches were presented in the literature [7,16]. In order to cope with challenges related to the use of non-overlapping cameras where changes of illumination may occur, visual descriptors that are independent on those fluctuations were proposed [1,2,10,11,13]. Moreover, methods for obtaining and for the usage of spatio-temporal dependencies were also developed. They employ various approaches like: particle filters [13], Bayesian networks [10], Markov chains [11], probability dispersion function [2] and others.

3   Used Method

As an input for the re-identification method the so-called observations are needed. The single observation contains 3 types of information: descriptors of object's visual features; location of the observed object; time of the observation. In order to compare and to match a pair of the observations, a measure of fitness has to be considered. A role of such a measure can be fulfilled with a probability of identity of the object represented by the pair of observations. It can be, in turn, formulated as the following Eq. 1:

$$P_i(O_A, O_B) = w_v \cdot P_v(O_A, O_B) + w_t \cdot P_t(O_A, O_B) + w_b \cdot P_b(O_A, O_B) \qquad (1)$$

where O_A, O_B is a pair of observed object, P_v, P_t and P_b are probabilities of identity for the given object which are based on visual features, spatio-temporal dependencies and object behavior patterns, respectively. Thus, w_v, w_t, w_b are weights of importance of particular probabilities ($w_v + w_t + w_b = 1$).

The modification of re-identification method is related to an incorporation of spatio-temporal dependencies and behavior model. Thus, the above formula need to be changed as follows:

$$P_i(O_A, O_B) = w_v \cdot P_v(O_A, O_B) + w_{tb} \cdot P_b t(O_A, O_B) \qquad (2)$$

where P_{tb} is probability of identity for the given object that is based on combined spatio-temporal dependencies and object behavior patterns.

3.1   Spatio-Temporal Dependencies Model

In order to describe spatio-temporal dependencies, time of transition is expressed with the probability density function. Using Expectation-Maximization (EM) algorithm the probability of transition time is modeled with Gaussian Mixture Model (more precisely three gaussians were used). Thus, temporal dependencies are described by the following formula:

$$p(t_{trans}) = \sum_{i=1}^{i=M} w_i \cdot N(t|\mu_i, \sigma_i) \qquad \sum_{i=1}^{i=M} w_i = 1$$

where: $p(t_{trans})$ express the probability of given time of transition $t_t rans$ between the particular pair of cameras; $N(t|\mu_i, \sigma_i)$ determines the value of normal distribution for the given time of transition; the parameters of the distribution are described with the mean value μ_i (mean transition time related to gaussian i) and standard deviation σ_i; w_i is the weight assigned to the particular gaussian; M is the number of gaussians in the GMM.

3.2   Behavior Pattern Modeling

The previous approach proposed by the authors is based on the idea of Pawlak's flow graphs[]. A transition between pairs of cameras can be described with a rule (IF conditions, THEN decision). All rules are organized into a graph which contains a knowledge about patterns of objects movements (called also a flow graph). Each rule is described with 3 parameters: strength, certainty and coverage [15] (see Eq. 3).

$$\sigma(x_i, y_j) = \frac{\varphi(x_i, y_j)}{\varphi(G)} \qquad \sigma(x_i) = \frac{\varphi(x_i)}{\varphi(G)} \qquad cer(x_i, y_j) = \frac{\sigma(x_i, y_j)}{\sigma(x_i)} \qquad (3)$$

where $\sigma(x_i, y_j)$ express the strength of rule and describes the rate of objects passing from location x in step i of the path to location y in step j; the number of objects passing from x_i to y_j is denoted as $\varphi(x_i, y_j)$, and the total number of paths in the input data is denoted as $\varphi(G)$; similarly, $\varphi(x_i)$ is the number of objects that passed location x in the step i; value of $cer(x_i, y_j)$ estimates the probability that the object which left location x in step i of the its path will appear in location y in step j. Therefore, the input statistical data necessary for building a flow graph is a set of objects' paths through the observed area. A single path is a sequence of locations (called also as steps) visited by the object.

Details related to the description, obtaining and using spatio-temporal dependencies and Pawlak's flow graphs in the re-identification are contained in the previous authors' publications [5,6,14].

3.3 Using Particle Filter

In order to combine premises from the topology graph and the flow graph, the proposed particle filter contains a number of particles related to possibilities of the object movement in the given location (according to [9]). The particle filter is updated as time of transition is elapsing. The formulas used to update particle filter for the given moment of time are presented below (Eq. 4):

$$m(y_{j,t_1}|x_{i,t_0}) = cer\,(x_i, y_j) \cdot p_{xy}\,(t = t_1) \tag{4}$$

$$p(y_{j,t_1}|x_{i,t_0}) = \frac{m(y_{j,t_1}|x_{i,t_0})}{\sum_{adj\_loc}} \tag{5}$$

where $cer\,(x_i, y_j)$ is certainty taken from the flow graph for the given pair of adjacent locations, $p_{xy}\,(t = t_1)$ is the value of probability destiny function for given moment of transition t_1 for particular adjacent locations (the transition is from x and y).

3.4 Decision Making

Previously, the non-modified method used weighting of three premises in order to obtain a probability of a particular object identity (see Eq. 1). Using the particle filter allows for a reduction of a number of summand for weighting (see Eq. 2). For each pair of observations the probability of object identity can be determined and pairs with the highest probability are chosen as those representing the transition between cameras of the real object.

4 Experiments and Results

Pilot tests of the system were performed. Before the experiments were made, the topology and the flow graph had to be built up as a part of the initialization procedure of the algorithm. Moreover, as modification of the re-identification method, the flow graph and the topology graph are combined into a particle filter. These graphs are presented in Fig. 1. Employing the equations presented in Sect. 3.1 the parameters of GMMs for the particular transitions are shown in Table 1. Moreover, the values of behavior pattern model are presented in Table 2 (the symbols are consistent with Sect. 3.2).

The prepared ground truth base contains only correctly detected transitions of objects through the whole observed area (cases of utterly correct tracking and detection for each single camera). The re-identification method (modified

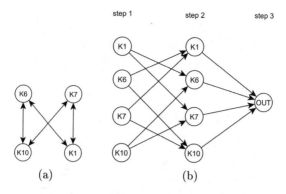

Fig. 1. Graphs used for estimation of probabilities P_t and P_b (See Eq. 1): (1(a)) topology graph; (1(b)) flow graph - vertex labeled as 'OUT' refers to the event when an object is leaving the observed area

Table 1. Parameters of GMM for the topology graph used for experiments (time is expressed in seconds)

Transition		w_1	μ_1	σ_1	w_2	μ_2	σ_2	w_3	μ_3	σ_3
From	To									
K1	K6	0.599706	55.6548	16.4292	0.166952	81.9792	12.2291	0.233343	67.5843	6.9635
K1	K7	0.501205	80.5872	3.1122	0.298830	73.9965	0.6657	0.199966	1.0034	89.0002
K6	K1	0.162216	69.3590	0.8711	0.166667	81.6667	1.5556	0.671117	56.1650	1.5556
K6	K10	0.281250	79.1111	2.9877	0.281250	79.1111	2.9877	0.437500	67.5714	2.1020
K7	K1	0.312233	82.3893	13.9951	0.322821	90.3258	6.7113	0.364946	85.2627	3.1618
K7	K10	0.357164	122.5990	18.6577	0.354433	96.3335	15.3770	0.288403	103.0196	2.5185
K10	K6	0.219122	78.9915	1.0032	0.345710	84.3249	1.5292	0.435169	91.8811	19.4572
K10	K7	0.342896	115.8483	24.7113	0.493019	100.1222	5.0633	0.164085	132.5366	6.2494

and non-modified) was applied to these pairs of observations. As a result the efficiency of the modified method can be compared with the previous one. FoVs of cameras used for the experiment are presented in Fig. 2.

Two hours of video data were analysed manually during the preparation of data necessary for initialization of the topology graph and the flow graph. In this period 76 object passed through the observed area and their timestamps and locations were determined. The next step was the preparation of the ground truth base. Firstly, other one hour fragments of video data from the cameras had to be analysed, in order to obtain location and timestamps of events related to

Table 2. Values of certainty (according to Eq. 3) for the flow graph presented in Fig. 1(b)

	Transition from x_1 to y_2							
	K1→K6	K1→K7	K6→K1	K6→K10	K7→K1	K7→K10	K10→K6	K10→K7
$cer(x_1, y_2)$	0.7059	0.2941	0.3600	0.6400	0.3913	0.6087	0.4286	0.5714

Fig. 2. FOVs of cameras and possible transitions between them (referred to as topology of camera network)

entering or exiting of the underpass. During this period 39 objects were observed. After that an automatic analysis was performed for each single camera, in order to remove those observed objects which were detected incorrectly or those which were not detected at all. Moreover, timestamps were corrected according to the time of detected events (entering or exiting of the underpass). As the result of this step 28 of real object remained in the ground truth base for the testing of the re-identification efficiency.

According to Eq. 1 particular weights were modified, in order to determine the best values of weights in reference to efficiency. The results are presented in Table 3.

The sum of all weights has to be equal to 1, therefore adding the third weight implicates a small modification in the calculation of the values of weights which is determined by the following formula (Eq. 6):

$$w_x = \frac{w'_x}{w'_v + w'_t + w'_b} \tag{6}$$

where $w'_x = w'_v, w'_t, w'_b$ and w_x is the value of the weight x after modification. It is to notice that if a single weight w'_x is equal to zero, the values of the rest of weights are exactly like those given in Table 3. In case of using only two types of measures of the probability of identity (that are related to visual features and to the topology graph) the best result was obtained for the following weights $w'_v = 0.5$ and $w_t = 0.5$. An addition of the third probability measure related to behaviour patterns causes a slight improvement of the following values $w'_v = 0.375$, $w'_t = 0.625$, $w'_v = 0.125$. The modification related to particle filtering caused no significant improvement of efficiency.

Table 3. Comparison of the number of correct re-identifications versus weights (for the particle filter method). In the ground-truth base there are 28 object transitions to be detected.

w_v'	w_b' .000 / w_t' 1	.125 / .875	.250 / .750	.375 / .625	.500 / .500	.625 / .375	.750 / .250	.875 / .125	1 / 0	w_{tb} 1	.875	.750	.625	.500	.375	.250	.125	0
.000	9	7	3	1	0	0	0	0	0	3								
.125	11	9	4	1	0	0	0	0	0		7							
.250	15	11	15	7	3	0	0	0	0			9						
.375	17	20	7	9	4	2	0	0	0				13					
.500	19	13	3	10	5	0	0	0	0					21				
.625	13	7	1	7	0	0	0	0	0						15			
.750	9	2	0	0	0	0	0	0	0							9		
.875	7	2	0	0	0	0	0	0	0								8	
1	4	1	0	0	0	0	0	0	0									4

5 Conclusions

In order to compare two versions of the re-identification algorithm an implementation was made and experiments were organized. The advantage of the particle filter application is related to the reduction of parameters (in this case weights) that need to be modified while adjusting the algorithm settings. The optimal weight values may be different for various types of cameras and for diversified spatio-temporal dependencies. The initialization process is needed before applying the proposed method to a new camera setup. Future work will be related to combining of all types of premises (visual, spatio-temporal and behavioral) in a single data structure, in order to avoid the weighting procedure, which is not longer necessary in this case.

Acknowledgements. This work has been partially funded by the Polish National Science Centre within the grant belonging to the program "Preludium" No. 2014/15/N/ST6/04905 entitled:"Methods for design of the camera network topology aimed to re-identification and tracking objects on the basis of behavior modeling with the flow graph".

References

1. Cheng, Y.S., Huang, C.M., Fu, L.C: Multiple people visual tracking in a multi-camera system for cluttered environments. In: 2006 IEEE/RSJ International Conference on Intelligent Robots and Systems, pp. 675–680 (2006)
2. Colombo, A., Orwell, J., Velastin, S.: Colour constancy techniques for re-recognition of pedestrians from multiple surveillance cameras. In: Workshop on Multi-camera and Multi-modal Sensor Fusion Algorithms and Applications, pp. 1–13 (2008)

3. Czyżewski, A., Szwoch, G., Dalka, P., Szczuko, P., Ciarkowski, A., Ellwart, D., Merta, T., Lopatka, K., Kulasek, Ł., Wolski, J.: Multi-stage video analysis framework. In: Weiyao, L. (ed.) Video Surveillance, chapter 9, pp. 145–171. Intech (2011)
4. Czyżewski, A., Dalka, P.: Moving object detection and tracking for the purpose of multimodal surveillance system in urban areas. In: Tsihrintzis, G., Virvou, M., Howlett, R., Jain, L. (eds.) New Directions in Intelligent Interactive Multimedia, Studies in Computational Intelligence, vol. 142, pp. 75–84. Springer, Heidelberg (2008)
5. Czyżewski, A., Lisowski, K.: Adaptive method of adjusting flowgraph for route reconstruction in video surveillance systems. Fundam. Inf. **127**(1–4), 561–576 (2013)
6. Czyzewski, A., Lisowski, K.: Employing flowgraphs for forward route reconstruction in video surveillance system. J. Intell. Inform. Syst. **43**(3), 521–535 (2014)
7. Dalka, P., Ellwart, D., Szwoch, G., Lisowski, K., Szczuko, P., Czyżewski, A.: Selection of visual descriptors for the purpose of multi-camera object re-identification. In: Feature Selection for Data and Pattern Recognition, pp. 263–303. Springer, Heidelberg (2015)
8. Dalka, P., Szwoch, G., Ciarkowski, A.: Distributed framework for visual event detection in parking lot area. In: Proceedings of 4th International Conference on Multimedia Communications, Services and Security, MCSS 2011, Krakow, Poland, 2–3 June 2011, pp. 37–45. Springer, Heidelberg (2011)
9. Doucet, A., Johansen, A.M.: A Tutorial on Particle Filtering and Smoothing: Fifteen years Later. Technical Report (2008). http://www.cs.ubc.ca/~arnaud/doucet_johansen_tutorialPF.pdf
10. Javed, O.: Appearance modeling for tracking in multiple non-overlapping cameras. In: IEEE International Conference on Computer Vision and Pattern Recognition, pp. 26–33 (2005)
11. Kim, H., Romberg, J., Wolf, W.: Multi-camera tracking on a graph using Markov chain monte carlo. In: Third ACM/IEEE International Conference on Distributed Smart Cameras, ICDSC 2009, pp. 1–8 (2009)
12. Kim, K., Chalidabhongse, T.H., Harwood, D., Davis, L.: Real-time foreground-background segmentation using codebook model. Real-Time Imag. **11**(3), 172–185 (2005)
13. Lev-Tov, A., Moses, Y.: Path recovery of a disappearing target in a large network of cameras. In: Proceedings of the Fourth ACM/IEEE International Conference on Distributed Smart Cameras, pp. 57–64, ICDSC 2010. ACM, New York (2010)
14. Lisowski, K., Czyzewski, A.: Complexity analysis of the pawlak's flowgraph extension for re-identification in multi-camera surveillance system. Multimed. Tools Appl. **75**, 1–17 (2015)
15. Pawlak, Z.: Rough sets and flow graphs. In: Slezak, D., Wang, G., Szczuka, M.S., Düntsch, I., Yao, Y. (eds.) RSFDGrC (1). Lecture Notes in Computer Science, vol. 3641, pp. 1–11. Springer (2005)
16. Radke, R.J.: A survey of distributed computer vision algorithms. In: Handbook of Ambient Intelligence and Smart Environments, pp. 35–55. Springer, Boston (2010)
17. Stauffer, C., Grimson, W.E.L.: Adaptive background mixture models for real-time tracking. In: CVPR, pp. 2246–2252. IEEE Computer Society (1999)
18. Szwoch, G.: Performance evaluation of the parallel codebook algorithm for background subtraction in video stream. In: Dziech, A., Czyzewski, A. (eds.) Multimedia Communications, Services and Security, Communications in Computer and Information Science, vol. 149, pp. 149–157. Springer, Heidelberg (2011)

Large LED Displays Panel Control Using Splitted PWM

Przemysław Mazurek[✉]

Department of Signal Processing and Multimedia Engineering, West Pomeranian
University of Technology, Szczecin, 26. Kwietnia 10 St., 71126 Szczecin, Poland
przemyslaw.mazurek@zut.edu.pl

Abstract. Multiplexed LED arrays are important for the design of large
scale displays used in indoor and outdoor applications. Novel brightness control technique - Splitted PWM is proposed in the paper. This
technique preserves turn–on time for LED that is not achieved in BCM
technique and reduces the number of comparisons required for light modulation. Overall system was implemented on xCORE 200 processor and
tested on 192×96 display. Memory organization is considered also, with
quad buffering of UDP/IP transmitted video stream.

Keywords: LED displays · PWM · Light modulation
Microcontrollers · Embedded systems

1 Introduction

Numerous display technologies are used nowadays. CRT display technology is
now obsolete and it is surpassed by LCD. Two techniques of backlights are used -
CFL and LED. There are three important problems of this technology - poor
black level, reduced brightness and limited size of screen. Alternative technology
based on LED exists but it is expensive solution and consume a lot of power [3,5].
Two approaches are used for LED display. First is based on design screen with
very small LEDs with good filling of panel so gaps between LEDs are very
small. Such panels are alternative to LCD applications: computer or TV screens
with rather small size. The second is based on LED modules that are connected
together for large and very large screens. Such displays are monochromatic,
bichromatic or RGB color. The distance between LEDs is large so area between
them is black or occupied by lenses/diffusers [4]. This last type of LED panels
is considered in this paper. Such panels are applied as information displays in
indoor and outdoor applications. They are used for different purposes, e.g. as a
passenger information displays, screens at mass events (e.g. concerts), billboards
replacements with commercials.

Large LED displays are not simple to design if the low–cost design is required.
The typical display uses modules with 16, 32, or 64 width or height (number of
pixels in horizontal or vertical directions). Different combinations of widths and

© Springer Nature Switzerland AG 2019
M. Choraś and R. S. Choraś (Eds.): IP&C 2018, AISC 892, pp. 87–95, 2019.
https://doi.org/10.1007/978-3-030-03658-4_11

heights are available for the design with cost effective fitting to the application. Different types of LEDs are used so different brightness ranges, pixels spacing, directional characteristics, color characteristics are achieved.

The number of manufactures is large, because most modules use similar hardware design and HUBxx interfaces (where xx is number related specific variant of interface and HUB75 is typical for RGB modules). Such modules are marked as P–modules, with attached digit (e.g. P4, P5), but modern marking is long alphanumeric marking that describe details of modules.

The main problem of P–modules is very simple design for the reduction of module costs, that influences on sophisticated control of the panel. There are additional problems like lack of standard, so in the most cases test and trial method is necessary during the development. It is related especially to timing parameters.

The example module is shown in Fig. 1.

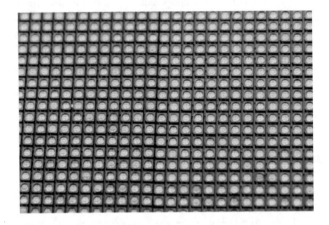

Fig. 1. LEDs of example P–module

1.1   Related Works

Control of P–modules could be based on PWM (Pulse Width Modulation). This is direct solution, that is very simple in concept but influences on design problems.

Alternative approach is based on BCM (Binary Code Modulation) also known as BAM (Bit Angle Modulation) [1]. This technique is more efficient due to low computation cost, but creates dynamic artifacts due to LEDs switching.

Different software and hardware solutions are available on market - FPGA (Field–Programmable Gate Array) and fast microcontrollers are used. Both variants have advantages and disadvantages. FPGA devices require external memory so a lot of pins are required for the connection to DRAM (Dynamic Random–Access Memory). Large pin count FPGA devices are expensive and high bandwidth is required. Fast microcontrollers may use internal or external memory for image buffer. The development cost is much higher for FPGA comparing to microcontrollers. The problem of different HUB interface variants increases development costs for both solutions.

Additional problem is the image delivery to display controller. Cost effective solution is Ethernet link.

1.2 Content and Contribution of the Paper

Proposed solution is based on high performance microcontroller: XMOS xCore 200 series. This microcontroller is multicore with 16 cores divided into two groups (tiles) [8]. Different variants of LED panels control were tested and a new modulation technique is proposed - Splitted PWM (SPWM). This modulation is designed for the omitting BCM problems and it offer quite low memory requirements. Multicore design of software allows the real–time processing image and display control as well as support of Fast Ethernet link.

Section 2 shows basic information about LED arrays. Section 3 is related to P–display control techniques. Section 4 shows results and provide discussion. Section 5 provide conclusions and further work.

2 Multiplexed LED Arrays

A few LED control techniques are possible for implementation. The naive solution is based on direct control every LED by single pin of microcontroller of FPGA. PWM signal applied for such pin (and LED) allows brightness changes. Small number of LEDs could be controlled using this technique but even small arrays require large pin count devices. There are 1024 LEDs assuming small 32×32 LED module so 1024 pins are required.

Alternative approach is based on intelligent LEDs with own IC (Integrated Circuit) [7]. This IC is addressed and has own PWM registers as well as communication serial interfaces. Three or four PWM registers are sufficient for control RGB and RGBW LEDs. The cost of single intelligent LED is problem. Such LEDs are daisy chained and chip should support regenerative interface, due to problem of digital signal passing over hundreds of devices and EMI (Electro–Magnetic Interferences).

Well known cost effective solution is multiplexed addressing of LED arrays [3]. Instead of control single LED a multiple LEDs are switched on at one time. Simplest solution assumes the control of N columns and M rows using N lines for columns and P for rows. Single row could be selected at one time, so the number of rows is:

$$M = 2^P. \tag{1}$$

Such multiplexed addressing requires $N + P$ lines. There are 37 lines required for 32×32 LED module.

P–modules use serial interface with four interface lines for column so the number of lines is reduced to 9 only. This interface uses shift register for with latches and output enable so data are transmitted and stored independently. Display is turned off by the output enable line. Different variants of this concept are applied in P–modules for the reduction of flickering using lines interleaving and zigzag. Multiple modules are daisy chained and the number of them is limited by numerous factors, like refresh, color resolution, timings of modules. The HUB interface is cost effective solution that uses TTL signals, but the lack of EMI protection in cables creates significant crosstalk interferences, so short cables are required. Maximal achievable clock speed is about 50 MHz for some modules, working at room temperatures. Different modules of the same series may have different timings, so clock frequency should be lower a few times.

The simplified schema of module is shown in Fig. 2.

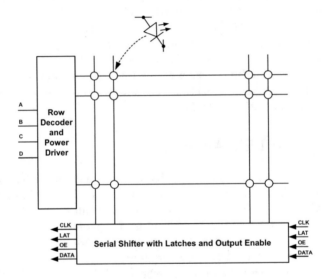

Fig. 2. Scheme of module

RGB signals are transmitted using independent data lines, but they share control signals.

3   Control Techniques of Multiplexed LED Array with Serial Interface

Simplest variant of LED control uses PWM. PWM allows brightness changes using counter and comparator. Counter could be common for all LEDs but comparator should be independent for each LED. An example 32×32 LEDs module requires 1024 hardware counters, so FPGA design is possible. Microcontrollers require different design with counter filled by values stored in the image buffer.

Larger displays require a lot of counters, so FPGA design should be based on counter filling in real–time, not individual counters. The size of the counter is dependent on desired brightness/color bit resolution. The RGB24 mode requires 8–bit counters for every color channel. Gamma curve correction requires larger resolutions e.g. 11–bit counters. The RGB24 mode is not required in numerous applications and simplified RGB15 or RGB16 modes are sufficient (using 5 or 6–bit counters).

The problem of multiplexed LED arrays is the conversion of RGB image to bit sequence required for HUB interface. Each LED is addressed 255 times per displayed frame. Flicker free mode requires more then 75 fps [6]. Simple 32×32 display requires $1024 \cdot 256 \cdot fps$ comparison of pixels values with counter. The number of operations is about 200M comparisons for very small display with single color. RGB display requires 600M comparisons. Memory wit large width and parallel comparators allows the processing of such data streams, especially using FPGA devices. Assuming 25 fps and repetition of frames allows preprocessing of image to interface bitstream. It allows the reduction of computation cost by factor 3 (for 25 fps of image data and 75 fps of refresh). The cost is the size of memory, that is quite low 32 kB and desired computation power for conversion of data to display bitstream. The cost is increased linearly for large panels. The number of comparisons is very large, but modern high speed microcontrollers supports SIMD (Single–Instruction Multiple–Data) processing [2], so 32–bit processor could support comparison processing for four independent values. The comparison operation requires data merging into bitstream so computation cost is large.

Alternative approach is BCM. This is more advanced LED control scheme. The idea of BCM is very interesting. There are no comparisons at all and the bitstream creation is more straightforward. Pixel value (single channel) is coded using 8–bits. Every bit has own weight $(2^7, 2^6, \cdots, 2^0)$ that corresponds to the LEDs turn on time directly. The processing of LED in PWM technique requires 256 operations for single pixel but BCM requires 8 operations only. BCM starts from MSB (Most Significant Bit) and turn on or off LED for specific time that is linear function of bit weight. Even simple microcontrollers could process using BCM.

BCM is good for ideal LED arrays, but real modules have transient problems. BCM gives dynamic problems of LED what is important for video displaying. Changes of brightness of single pixel are not smooth.

Alternative approach is proposed in this paper using Split PWM (SPWM). SPWM uses intermediate solution between PWM and BCM. PWM turns on LED for some period and due to multiplexing this LED is driven by fixed frequency modulated signal. BCM frequency modulation depends on pixel value. SPWM uses fixed frequency modulated signal but the number of steps for processing of LED frame cycle is between PWM and BCM. Exemplary signals for control of single LED are depicted in Fig. 3.

SPWM method introduced in this paper divides binary input value into two parts (MSP: Most Significant Part and LSP: Less Significant Part). The number

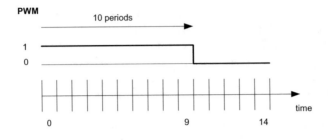

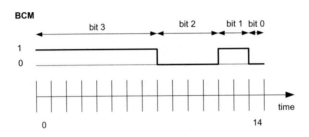

Fig. 3. PWM and BCM control (example of decimal value 10 coding for 4–bit values)

of bits in RGB15 mode is 5, so MSP part uses 2 bits and LSP part uses 3 bits. Coding of MSP part requires 3 time slots of PWM (no. 0, 1, 2), filled by values: 000b (MSP = 00b), 001b (MSP = 01b), 011b (MSP = 10b) or 111b (MSP = 11b). The LSP part requires additional time slot (no. 3) filled by PWM sequence in reversed order: 0000000b (MSP = 000b), 1000000b (MSP = 001b), 1100000b (MSP = 010b), ⋯, 1111111b (MSP = 111b). Reversion in LSP part gives the continuity of LEDs turn–on time. The time for single MSP time slot is equal to overall LSP time plus one.

The most important difference between BCM and SPWM is the lack of additional modulations forced by value for SPWM. The continuous period for turn on time is preserved as in PWM, but without large scale comparison. Visual results are comparable to PWM.

4   Results and Discussion

The controller was tested with different types of LED modules. The main application uses 192×96 resolution using 18 of P5 modules (32×32 pixels). The brightness of panel is constant and highest achievable for this module. Controller uses XMOS XEF216–512–TQ128 processor that supports two 256 kB SRAM (Static Random Access Memory) memories for data and program code [10].

The organization of LED chains is selected for the maximization of display size and the reduction of clock speed due to EMI. Overall display of 192 × 96 size is organized as a set of 6 chains of modules (Fig. 4).

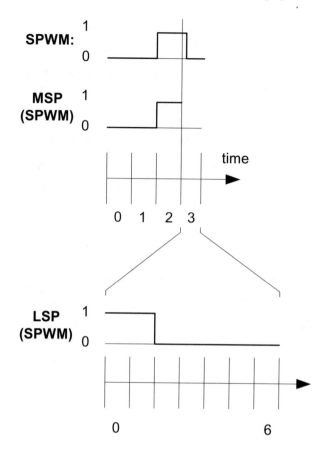

Fig. 4. SPWM control (example of decimal value 10 coding for 5–bit values)

Tile 0 is responsible for processing four chains with 96×32 resolution (A and B chains). Tile 1 is responsible for two chains with 96×32 resolution (C chains). Such design allows the reduction of line lengths for EMI suppression and the controller is located in a middle of panel. Such configuration is possible due to large number of I/O pins of selected processor.

The design is hardware flexible, so different configurations of display (overall size, modules size, types, timing) is possible. The tilling of multiple displays is possible, so larger displays are obtained by mechanical merging of them. Ethernet connectivity allows the signal distribution using multiport switch or switches.

Tile 1 is responsible for the UDP/IP stack processing. xCORE 200 processors support Ethernet up to Gigabit Ethernet using dedicated hardware, but some cores of Tile 1 must be assigned for partial processing of data and support of software UDP/IP stack [9]. PHY, MII and SMI drivers as well as ICMP and UDP server uses 5 cores of Tile 1. Data transmission of RGB565 at 25 fps requires less then 8 Mbit/s for 192×96 resolution.

Image data are received by Tile 1 and are available for all cores of this tile by shared memory. Three cores related to bitstream conversion and transmission to 1/3 of display are used in Tile 1.

Image data are transferred to Tile 0 from Tile 1 because both of them have independent shared memories. Six of eight cores of Tile 0 are used for bitstream conversion and transmission to 2/3 of display.

Quad buffering of data is achieved (Fig. 5). First buffer is filled by incoming data. Second buffer is filled by image data and transmitted to Tile 0. Tile 0 has own receiving image buffer. Data are converted to third buffer in a SPWM format. The last buffer is used for sending data to multiple HUB interfaces. Two last buffers are swapped dynamically for protection against artifacts. Synchronization of processes is hardware supported by XMOS processor [8].

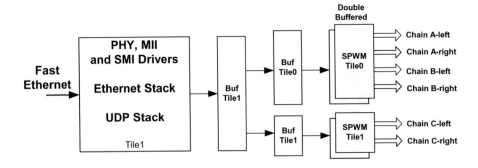

Fig. 5. Memory buffers

HUB interface uses software processed I/O. XMOS support hardware serializers/deserializers (SERDES), but some issues related to synchronous operation serializers occur, so software version is used as a fix. It is possible due to high speed cores. Overall code is written using xC (parallel version of C, with some extensions related to XMOS processors) using xTime Composer Studio.

Final design uses most of SRAM memory and 75 fps is achieved (refresh rate) for 25 fps image data rate.

5   Conclusions and Further Work

All presented techniques (PWM, BCM and SPWM) were implemented on xCORE processor, but PWM was quickly dropped from development due to large scale display problem support due to lack computation power. The design of SPWM was influenced by BCM idea. The selection of MSP and LSP values was based on evaluation of memory requirements and optimal variant was selected.

The reduction of flickering is possible by the selection of modulation technique. Further work will be related to investigation of another modulation techniques for the flickering. SPWM flickering is similar to PWM, but reduced flickering is observed in BCM.

References

1. Led dimming using binary code modulation. www.batsocks.co.uk/readme/art_bcm_1.htm
2. Cyper, R., Sanz, J.: The SIMD Model of Parallel Computation. Springer, New York (2011)
3. Lv, X., Loo, K., Lai, Y., Chi, K.T.: Energy efficient led driving system for large-scale video display panel. In: 39th Annual Conference of the IEEE Industrial Electronics Society, IECON 2013, pp. 6063–6068. IEEE (2013)
4. Motlagh, A.A.N., Hashemi, B.M.R.: Minimal viewing distance calculation in led display panels. J. Display Technol. **6**(12), 620–624 (2010)
5. Shyu, J.C., Hsu, K.W., Yang, K.S., Wang, C.C.: Performance of the led array panel in a confined enclosure. In: 5th International Microsystems Packaging Assembly and Circuits Technology Conference (IMPACT), pp. 1–4. IEEE (2010)
6. Wilkins, A., Veitch, J., Lehman, B.: Led lighting flicker and potential health concerns: IEEE standard PAR1789 update. In: Energy Conversion Congress and Exposition (ECCE), pp. 171–178. IEEE (2010)
7. Worldsemi: WS2812 Intelligent contol LED integrated light source. Worldsemi (2018)
8. XMOS: XMOS Programming Guide. XMOS (2015)
9. XMOS: Ethernet MAC library 3.3.0. XMOS (2017)
10. XMOS: XEF212-512-TQ128 Datasheet. XMOS (2017)

Air-Gap Data Transmission Using Backlight Modulation of Screen

Dawid Bąk$^{(\boxtimes)}$ and Przemyslaw Mazurek

Faculty of Electrical Engineering, Department of Signal Processing and Multimedia
Engineering, West Pomeranian University of Technology, Szczecin,
26 Kwietnia 10 St., 71-126 Szczecin, Poland
{dawid.bak,przemyslaw.mazurek}@zut.edu.pl

Abstract. Novel technique for data transmission from air–gap secured computer is considered in this paper. Backlight modulation of screen using BFSK allows data transmission that is not visible for human. The application of digital camera equipped and telescope allows data recovery during the lack of the user's activity. Demodulation scheme with automatic selection of demodulation filters is presented. Different configuration of data transmission parameters and acquisition hardware were tested.

Keywords: Air–gap transmission · BFSK · Image processing
Digital demodulation · Network security

1   Introduction

Modern problem of digital communication technologies is the risk of data theft. Isolated computer systems (isolated computers and isolated networks) are applied for the reduction of this risk. These systems are not connected to external networks at all and they are known as an air–gap systems. The lack of communication interfaces is used sometimes for further risk reduction and the protection of highly valuable data. It is illusory situation, because data leakage is possible using numerous hardware and software techniques applied for such computers using unidirectional data links. These data links are unconventional and should be hidden for regular security audits.

One important problem is the technique of computer infection by software that creates such link. This topic is outside of this paper scope.

The technique proposed in this paper uses unidirectional link based on the computer screen modulation using backlight changes. These changes could be invisible to human operator but they could be observed using camera or telescope equipped with camera. Small fluctuations are extracted from video image and processed by image processing as well as digital signal demodulation algorithms. Data could be received using real–time processing or off–line using acquired video image.

© Springer Nature Switzerland AG 2019
M. Choraś and R. S. Choraś (Eds.): IP&C 2018, AISC 892, pp. 96–103, 2019.
https://doi.org/10.1007/978-3-030-03658-4_12

One of the most important properties of proposed technique is the security vulnerability related to necessary privileges. Full access to system is not required, because user privileges are sufficient due to the lack of backlight changes protection.

1.1 Related Work

There are numerous air–gap transmission techniques, other techniques uses different physical mediums or computer properties for data transmission. One of the techniques is QR code embedding in images, that is know as a VisiSploit [2]. Another technique is xLED that is destined for the data transmission using light modulation of LEDs that are installed in network routers [8]. Similar technique is LED–it–GO that uses hard disk LED activity [7]. Other similar known techniques are aIR-Jumper [5] where the authors used a source of infrared light as a medium of hidden transmission. There are also several techniques that use an electromagnetic field as a transmission medium like USBee [4], AirHopper [3] and GSMem [6]. The previous work of authors of this paper is based on the modulation of brightness of screen [1].

2 The Idea of the Backlight Modulation Method

The main idea of proposed technique is the modulation of screen backlight. This modulation gives amplitude changes that could be not visible for human. In order to transmit data, computer should be infected.

2.1 Signal Generation

The following signal generator uses BASH script [11]. This code is presented as the illustration of the technique and shows security problem - data transmission does not require administrator privileges. Real software for data transmission should be written using more secure techniques for the protection against security audits.

```
#!/bin/bash
freq=0.5
del=4
l=$(expr length $1)
for ((i=1;i<=$1;i++));do
                input=$(expr substr $1 $i 1)
                if [ $input -eq 0 ];then
                   f=$freq
                   d=$del
                else
                    f=$(echo -n 0;bc <<< "scale=2;$freq/2")
                    d=$(($del * 2)
                fi
                t=0
```

```
while [ $t -ne $d ];do
    xbacklight -5
    sleep $f
    xbacklight +5
    sleep $f
    t=$(($t + 1))
done
done
```

The input of script is the bit sequence and 'xbacklight' program is used for backlight changes [10] for Linux operating system. BFSK modulation is applied [9] for data transmission. Two modulation frequencies are used f_0 and $f_1 = \frac{f_0}{2}$) for data transmission of '0' and '1' bit values respectively. The time for the transmission of bit is configured additionally. The transmission of '0' gives two times higher modulation of light comparing to '1' data transmission. BFSK is simple modulation scheme and is well protected against distortions, e.g. surrounding light changes.

2.2 Signal Acquisition

The technique of data acquisition is similar to the previous work [1] and is shown in Fig. 1. Infected computer change screen backlight and this screen is observed by camera. Direct visibility of the screen or part of screen is required. Two configurations of acquisition system were tested - cheap web camera with low resolution and digital camera attached to the telescope (Table 1).

Table 1. Devices

	Webcam	Digital camera
Model	Titanum TC102	OLYMPUS SZ-11
Resolution	640 × 480	1280 × 720
FPS	6	30
Codec	MPEG-2	MPEG-4 (H.264)
Additional optics	–	36050 Telescope

The example of computer screen is shown in Fig. 2. Different distances where tested and for webcam maximal distance is about 0.5 m. The telescope with camera gives correct results at 30 m distance. The low distance for webcam is related to the processing of full image from camera. The cropping of video stream could extend the distance.

2.3 Signal Demodulation

The signal demodulation code is written in Matlab and the demodulation scheme is shown in Fig. 3. One of the most important properties is automatic detection

Fig. 1. Data acquisition method

Fig. 2. Image recorded using a digital camera

of modulation frequencies. It is necessary because the modulation is based on computer timers that could be fixed but not known. The demodulation scheme is:

1. Reading of input video file,
2. Removal of constant signal (DC),
3. Removal of noise from signal, using zero–phase low–pass filter,
4. Estimation of center frequencies and bandwidth for band–pass filters using filtered spectrum (Fig. 3),
5. Band–pass filtering of signal using 4'th order filters designed in previous step (Fig. 4),
6. Envelope detection of signal using rectification and low–pass filters (Fig. 4),
7. Binary stream recovery using envelopes comparison (Fig. 4).

This demodulation uses off–line processing due to nonlinear zero–phase filter, but another type (linear) filter could be designed for this application. The estimation of center frequencies is based on FFT spectrum. This spectrum is noised, due to modulation artifacts, because smooth backlight switching is not possible. Additional spectrum noises are the results of surrounding lights. The vector spectrum amplitudes is filtered using additional zero–phase filter and three peaks are observed. The first peak is located at DC, second and third peaks are related to both modulation frequencies. The estimation of peak positions is based on local maximum finding. Achieved frequencies (lower f_1 and higher f_1) are used to the calculation of half bandwidth (HB) using the following formula:

$$HB = \frac{f_0 - f_1}{2},\tag{1}$$

so the lower band–pass filter has the following frequencies:

$$f_1^{Lo} = f_1 - HB,\tag{2}$$

$$f_1^{Hi} = f_1 + HB,\tag{3}$$

and higher band–pass filter has the following frequencies:

$$f_0^{Lo} = f_1^{Hi} = f_0 - HB,\tag{4}$$

$$f_0^{Hi} = f_0 + HB.\tag{5}$$

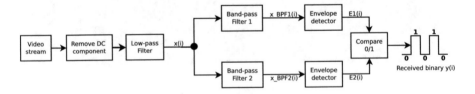

Fig. 3. Scheme of signal demodulation

3   Results

3.1   Example Case

The distance was about 30 m between computer and intruder equipped with the telescope and camera. Transmitted bit sequence is '0101011010101'.

The signal spectrum and amplitude characteristics of automatically designed filters are shown in Fig. 4.

The example of signal processing results for some parts of algorithm are shown in Fig. 5.

3.2   Tests

A few results for tests are presented in Table 2 for different setups.

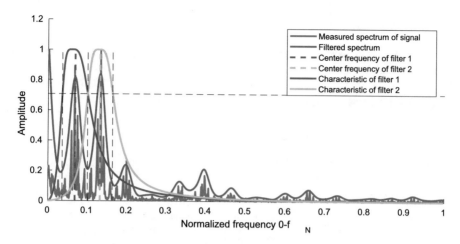

Fig. 4. Signal spectrum and designed filters

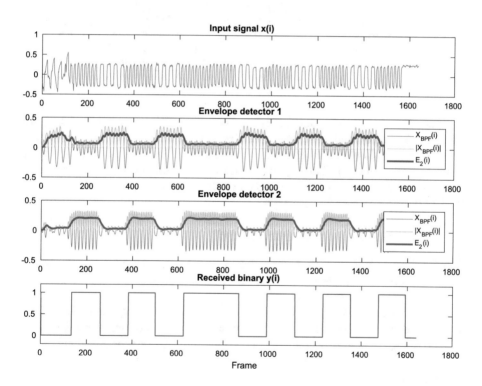

Fig. 5. Filtered signal and received binary data

4   Discussion

Table 2 shows approximate results, because there are numerous additional environmental conditions that could influence the process. Main result is the long distance attack with the use of telescope. The distance about 30 m is not maximal, but shows the attack possibility from another building.

Table 2. Test results with webcam

Device	Backlight difference	Period [s]	Camera distance [cm]	Transmitted sequence	Received sequence	Bit error rate
Webcam	10	8	40	0101011010101	0101011010101	0%
	10	8	40	0101011010101	0101011010101	0%
	4	8	40	0101011010101	0101011010101	0%
	4	4	40	0101011010101	0101011010101	0%
	2	4	40	0101011010101	0100011010101	7%
	4	4	55	0101011010101	0101011010101	0%
	4	4	100	0101011010101	0100000000000	46%
Camera with telescope	10	8	500	0101011010101	0101011010101	0%
	10	8	3200	0101011010101	0101011010101	0%
	4	8	500	0101011010101	0101011010101	0%
	4	4	3200	0101011010101	0101011010101	0%

The effective transmission speed is very low. The increasing of bit rate requires higher backlight changes that are visible if the value is greater then 10. Experimental evaluation gives the optimal time for bit about 4 s and backlight changes at level 4. The achieved data transmission speed is 0.25 bps. It is enough for data transmission of e.g. passwords. Assuming 64–bit password there are less then 5 min for the reception.

5   Conclusions and Future Work

Proposed technique is efficient for small data sizes transmission at moderated distances. This technique is valuable for transmission of passwords, PINs, etc. from air–gap secured systems.

The application of standard tools that are available on typical Linux/Unix operating systems is possible [12]. No special privileges are necessary for transmission using backlight modulation. The only one requirement is the turned on monitor. This technique could be applied for the period without user activity, because user's actions disturbs overall screen brightness.

Further work will be related to the estimation of the attack depending on light conditions.

Acknowledgment. This work is supported by the UE EFRR ZPORR project Z/2.32/I/1.3.1/267/05 "Szczecin University of Technology – Research and Education Center of Modern Multimedia Technologies" (Poland).

References

1. Bak, D., Mazurek, P.: Air-gap data transmission using screen brightness modulation. In: International Interdisciplinary Ph.D. Workshop (IIPhDW), pp. 147–150 (2018)
2. Guri, M., Hasson, O., Kedma, G., Elovici, Y.: An optical covert-channel to leak data through an air-gap. In: 14th Annual Conference on Privacy, Security and Trust (PST), pp. 642–649 (2016)
3. Guri, M., Kedma, G., Kachlon, A., Elovici, Y.: Airhopper: bridging the air-gap between isolated networks and mobile phones using radio frequencies. In: 9th International Conference on Malicious and Unwanted Software: The Americas (MALWARE), pp. 58–67 (2014)
4. Guri, M., Monitz, M., Elovici, Y.: USBee: air-gap covert-channel via electromagnetic emission from USB. In: 14th Annual Conference on Privacy, Security and Trust (PST), pp. 264–268, December 2016
5. Guri, M., Bykhovsky, D., Elovici, Y.: Air-jumper: covert air-gap exfiltration/infiltration via security cameras & infrared (IR). CoRR abs/1709.05742 (2017)
6. Guri, M., Kachlon, A., Hasson, O., Kedma, G., Mirsky, Y., Elovici, Y.: GSMem: data exfiltration from air-gapped computers over GSM frequencies. In: 24th USENIX Security Symposium (USENIX Security 2015), pp. 849–864. USENIX Association, Washington, D.C. (2015). https://www.usenix.org/conference/usenixsecurity15/technical-sessions/presentation/guri
7. Guri, M., Zadov, B., Atias, E., Elovici, Y.: LED-it-GO: leaking (a lot of) data from air-gapped computers via the (small) hard drive LED. CoRR abs/1702.06715 (2017). http://arxiv.org/abs/1702.06715
8. Guri, M., Zadov, B., Daidakulov, A., Elovici, Y.: xLED: covert data exfiltration from air-gapped networks via router LEDs. CoRR abs/1706.01140 (2017)
9. Haykin, S.S.: Digital Communications. Wiley, New Delhi (1988). https://books.google.pl/books?id=lPiO8J6VC5YC
10. Packard, K.: xbacklight(1) Linux User's Manual, July 2013
11. Robbins, A., Beebe, N.: Classic Shell Scripting: Hidden Commands that Unlock the Power of Unix. O'Reilly Media, North Miami Beach (2005). https://books.google.pl/books?id=jO-iKwPRX0QC
12. Tanenbaum, A.S., Bos, H.: Modern Operating Systems, 4th edn. Prentice Hall Press, Upper Saddle River (2014)

Using Different Information Channels for Affect-Aware Video Games - A Case Study

Mariusz Szwoch and Wioleta Szwoch[(✉)]

Gdansk University of Technology, Gdańsk, Poland
{szwoch,wszwoch}@eti.pg.edu.pl

Abstract. This paper presents the problem of creating affect-aware video games that use different information channels, such as image, video, physiological signals, input devices, and player's behaviour, for emotion recognition. Presented case studies of three affect-aware games show certain conditions and limitations for using specific signals to recognize emotions and lead to interesting conclusions.

1 Introduction

Affective computing studies how to automatically recognize, interpret, and process human emotions based on analysis of available sensory data [1]. *Affect-aware* applications take into account users' emotions to change their behaviour and better suit the users' needs. Such *affective interventions* distinct them from typical, non-affective applications. Video games seem to be one of the most natural affect-aware applications as they are based in much extent on player's emotions. Due to high competition in the video games market new approaches are still being tried to attract players' attention and tie them up for longer time with a given title.

Although the idea of creating affect-aware video games appeared almost two decades ago, so far only a few productions have been created with an option of reacting to the emotions of the player. One of the reasons was certainly problems with the effective recognition of emotions, and on the other hand, quite similar effect could be achieved in video games by using dynamic difficulty adjustment (DDA) or just the initial selection of the difficulty level by the player. Only in recent years, the development of methods for emotion recognition has allowed to obtain more accurate results.

According to many researchers, it is best for the play experience, if the player stays in the so called *flow channel* and the play experience follows a track more like in Fig. 1 [2]. Considering the player's emotions will make it much easier and more efficient to keep them in the flow channel using proper affective interventions.

Many researchers use face as the primary and sometimes the only source of information, recognizing facial expressions from static images and video recordings [3]. However, it is possible to use other expressions of human emotions like

© Springer Nature Switzerland AG 2019
M. Choraś and R. S. Choraś (Eds.): IP&C 2018, AISC 892, pp. 104–113, 2019.
https://doi.org/10.1007/978-3-030-03658-4_13

the speech, pose of the body, gestures or physiological signals [4,5]. Also, a great many machine learning algorithms have already been applied in the task of emotion recognition, like SVM, decision trees, linear discriminant analysis, Bayesian networks, naive Bayes, neural networks etc. [6,7]. This paper raises the problem of choosing different information channels for recognizing or estimating the player's emotions. Section 2 describes most popular of these input channels like image, video, bio-signals etc. Section 3 presents case study of three affect-aware video games that use different signals for emotion estimation. Section 4 shows results and conclusions from this work.

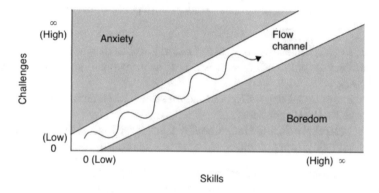

Fig. 1. The concept of keeping the player's interest in gameplay with the development of their skills by increasing the difficulty of the game [2]

2   Emotion Recognition for Affect-Aware Video Games

Although considering player's emotions in video games seems to be the natural direction of their evolution, no formalized methodologies or tools for their development have been proposed for a long time . The first such proposition was the AFFINT process [8] that lists, systematizes and organizes all activities and factors that should be taken into account during development of affect-aware video games. It formulates a 10-step process that lets us find a compromise between the three most important aspects of affective applications: system characteristics (goals, functionality and interface metaphor), knowledge of affective phenomena, and available affective computing solutions that are applicable in the system context.

According to AFFINT, one of the important elements of the design of affect-aware applications is the selection of input channels that provide information about the player's emotions. Unfortunately, there are currently no perfect methods for estimating the player's emotions. All proposed approaches are based on the choice of one or several information channels and provide a certain compromise between the user's convenience and measurement accuracy. The fact is that in the commercial software the method of obtaining data to recognize emotions should be perceived by the player as little as possible, even at the expense of the accuracy of estimating emotions.

Nowadays, this limits most reliable methods using physiological signals, such as electrocardiography (ECG), electroencephalography (EEG) but also, to some extent, many other methods, e.g. very popular recognition of facial expressions, which requires the player to face the camera all the time during the gameplay. Other limitations are for example the price and low popularity of different input or monitoring devices that might be used to estimate emotions, such as thermal cameras, RGB-D sensors, EEG brain monitors etc. That is why most researchers focus on three following solutions that are presented in more detail in the following subsections:

- heart rate (HR) measurement using smartwatches or smart bands that have become very popular lately and are now accurate enough and not intrusive for players;
- recognition of facial expressions based on standard cameras that are available in laptops and smartphones;
- player's behaviour during the gameplay that cover both their success rate and the way of using input devices, such as the keyboard, mouse etc.

Another problem is the selection of a set of recognized emotions and their representation. Although, people can express a huge range of different emotions, only six basic emotions distinguished by Ekman (joy, fear, anger, surprise, disgust and sadness) [9] are most often considered during development of affective applications. Moreover, in video games this subset is usually more limited to usually two emotions lying on the same axis of the PAD (pleasure, arousal, dominance) [10] model and the third neutral state. This is because some emotions rarely occur at the time of playing (e.g. disgust) and do not have too much influence on the planned affective interventions as most often game developers want to just detect the player's activity or passivity, frustration or boredom, happiness or sadness etc. It is also clear that the detection of a limited subset of emotions usually gives better results, which is important.

Yet another problem is to design the proper affective interventions as a reaction to the detected emotional state of the player. In most cases, when gameplay challenges become too hard for the player, the affective intervention reduce their difficulty, thus implementing postulates of the DDA approach. Of course, one can also imagine the opposite reaction, when the gameplay becomes more difficult for the more stressed the player, which could be used to teach the players to control their emotions.

2.1 Video Channel

Video is a very useful, valuable, and informative channel for emotion recognition because people tend to express their emotions in a visual way. It is also absolutely non-intrusive and video cameras are cheap, popular and available on most PC and portable platforms. Although, different body language signals can be used for emotion recognition, including gestures, posture, hand and body movements, the most common approach is based on automatic analysis of facial expressions, which is also the most developed field in affective computing [3]. Some of the existing approaches are based on detection of facial characteristic points like eyes, elbows, nose, mouth etc. using different heuristics and algorithms like texture matching, edge detection, and others. Landmarks' movements within the face area can be described by Facial Action Coding System (FACS) and finally map and classify different configurations (e.g. raised eyebrows) as different emotions [9].

Because emotion recognition based on facial expressions is very popular among researchers, many free and commercial libraries offering this functionality have been developed. The effectiveness of these tools is relatively high, especially when recognizing emotions associated with clear and unambiguous facial expressions, such as joy (smile) or surprise. Unfortunately, recognition of other emotions is usually less effective. An important drawback of most algorithms is their high sensitivity causing frequent result changing, which can most often be reduced by low-pass filtering in the time domain. Some disadvantage of this approach (as others based on video input) is that the whole player's face has to be observed in good and stable lighting conditions which is not always possible, as many players play in a dark room. Another problem is the need to ensure proper face orientation in relation to the camera and not too long distance, which is a problem for console players.

2.2 Bio-signals

Bio-signals can be very useful for emotion recognition [11] as they are to a large extent independent of human will, and reflect our real emotions, not the expressed ones. There are many physiological signals, but for emotion recognition the most common are: HR, skin conductance (SC), muscle tension (MT), ECG, EEG, etc. Psychophysiology shows that there is an evident correlation between bio-signals and arousal/valence of human emotions [12]. For example, SC changes, if the skin is sweaty, which is often related to the stress situations, thus can be used as an indicator of emotion arousal. Unfortunately, it demands usage of some electrodes often placed on a finger or shoulder, which is rather intrusive for the player. In turn, EEG can provide very useful information about human's behaviour and emotions but it demands electrodes located on the scalp being also intrusive. One of the least intrusive bio-signals is HR that can be measured by smart bands or smart watches that can be normally worn on the wrist. HR evidently rises in stress situations considering no physical effort is involved and can be used for player's arousal estimation.

One of the biggest problems of using bio-signals is that they are very individual, which significantly hinders their interpretation and creation of general algorithms. In practice, most algorithms that use them assume a certain period classifier's pre-training, so that it adapts to a specific user. Physiological signals are also heavily dependent on a variety of internal and external factors such as human health, physical condition, temperature etc. Another problem is that the sensors for signal measurement are often intrusive or even invasive for the users. For medical purposes it is not critical but for real-life applications it is not acceptable. Thus, many researchers focus just on wearable devices that measure HR signal.

2.3   Input Devices and the Player's Behaviour

Video games offer the player a set of input devices that can be used to control the actions during the gameplay. Standard devices include a mouse, keyboard and more often nowadays touch screens. In a natural way the player's emotions are transferred to the way of using these devices, which also has become the subject of research [13]. There are many different parameters that can be investigated in the time or frequency domain such as duration and frequency of single keystroke or their sequences, the speed or number of pauses of the mouse or cursor movement, the force of pressing the touch screen, and others.

Different SVM classifiers, neural networks, binary tree, C4.5 tree, Bayesian network, and statistical analysis are most often used for emotion recognition based on such parameters. The research shows that there is a correlation between player's emotions and certain features distinguished for mouse movement and keyboard typing. For example, [14] has shown that there is a correlation between arousal and mouse movement precision and smoothness. It is obvious that the parameters associated with the keyboard and mouse are closely dependent on the type of application used by the user. The way of using input devices is also largely individual, which often requires an additional phase of learning for a particular player.

There are also other features that can be used for the estimation of player's emotions. They are generally based on player's behaviour during the gameplay. For most players, success in the game translates into their good mood and *vice versa*. This assumption underpins the DDA mechanism in many games. But, this inference can also be used to estimate the player's emotions leading to more complex affective interventions during the gameplay. The most important advantage of this approach is independence from additional external devices, while the big disadvantage is the need to modify the core mechanics of the game. This drawback may be eliminated by moving the analysis of emotions beyond the game to the external module, which was proposed in [15].

3   Emotions in Video Games - Case Study

In this research, three most important input channels are considered including physiological signals, optical channel, and behavioural characteristic of the player. In the first case, player's emotion is recognized based on HR signal measured by a smartwatch. In the next study, sample FERTcl library is used to recognize player's emotions based on video input from a standard camera. Finally, a sample algorithm of determining players' emotions based on their behaviour during the gameplay was used. Although, this approach is not exhaustive, it allows, however, to detect and characterize the most important features associated with each channel of information under the examination.

3.1 Emotion Recognition Using Bio-signals

Teeth defense (TD) is a tower defense game, whose educational objective is teaching children in early school age the fundamentals of oral hygiene [16]. The enemies in the game are bacteria leading to tooth decay, food remnants and tartar, while the defense towers can be made using toothbrushes, toothpaste, water irrigators, and a visit to a dentist (Fig. 2, leftmost). The win condition is to defeat all waves of the incoming enemies before they reach the player's base. In the case of educational games, the most undesirable emotional states are boredom and excessive stress caused by too easy or too difficult gameplay. Because younger players may have negative associations with dentistry and oral hygiene, which may further aggravate negative emotions, the optimal mechanism is to reduce the impact of these emotional states and keep the player in a focused state.

When the player's stress is detected, the emotional feedback in the game extends the time between consecutive waves of opponents, slows the opponents down, weakens them, and changes the colour scheme to a calmer one. Conversely, when player's boredom is detected, the game's colour scheme changes to more stressful reddish, the number of enemies increases, they are faster and stronger.

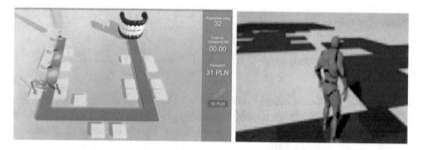

Fig. 2. Sample screens of Teeth defense and Ejcom (Crazy platforms level) video games

The emotion recognition of these three emotional states is based on player's HR measured by the Huawei Watch whose accuracy is enough sufficient. The device communicates with the phone and sends the heart rate of the player to the detection module. At the beginning of the game, the base heart rate is set for each individual user. Next, a range $\pm 7\%$ of the base HR is set, in which the user can be considered to be still focused on the gameplay. Exceeding the upper boundary of the division is interpreted as the appearance of stress, and descent below the bottom as a symptom of boredom. The detected emotional state is sent to the computer on which the game is being played and where the module of affective intervention is located.

3.2 Emotion Recognition Using Video Input

The logical-arcade educational game Ejcom consists of two levels: the Maze and Crazy platforms [17]. The first level assumes the development of logical thinking by memorizing and finding a route to the maze exit, while the second is to develop the player's motor skills, coordination and ability to make quick decisions. In the Maze level, the player has to control the movement of the walls in order to get to the exit. In contrast, in the Crazy platforms, the player goes to the exit over moving and disappearing elements (Fig. 2, rightmost).

The game uses FERTcl library, developed at Gdansk University of Technology, that recognizes players' emotional state based on their face from a standard camera. For each frame the probabilities of all six Ekman emotions and the neutral state are estimated. As the game's affective intervention is triggered after detection of the player's boredom or frustration, simple linear mapping has been experimentally determined. When calculating the player's frustration, two emotions are taken into account: anger and sadness. The player's boredom is calculated on the basis of the values of sadness, joy, anger and the neutral state.

If the value of boredom or frustration exceeds a certain threshold, certain affective intervention is taken. In the case when the player begins to get bored, the game becomes more difficult: in the maze, the time runs faster, the platforms move at a higher speed and in an unpredictable way, the colour of the platforms also changes. Conversely, when player's frustration is recognized, the difficulty of the game is reduced: the pace slows down and platforms move in more predictable direction.

3.3 Emotion Recognition Using Input Devices and Player's Behaviour

Wombat is a 3D platform affect-aware game with role playing elements [18]. The game is kept in a fairy tale convention and its goal is to defeat all of the enemies and save the princess. The game distinguishes three emotional states of the player, which are nervousness, neutral and boredom. Evaluation of the player's emotions takes into account the usage of the mouse and keyboard. To calculate the user's nervousness and boredom the application measures the mouse speed and the number of keys pressed during the last one second and, additionally, simultaneously. Experimentally determined thresholds indicate the detection of one of the interesting states. Two other classifiers detect undesirable states based on both: damages, life points and maximal life points as well as avatars' collisions with the elements of the game world. These four components are added with weights and the player's nervousness or boredom are calculated.

According to these two states, the game takes an appropriate affective intervention. If the user seems to be bored, the world becomes less friendly - the colours of the world transform to dark, additional enemies appear and less useful items can be found (Fig. 3, leftmost). Conversely, when the player gets nervous, the game converts to a friendly and easier one - the colours come back to cheerful and bright, some enemies disappear and more items can be found (Fig. 3, rightmost).

Fig. 3. Sample screen of two modes of the game

4   Discussion and Conclusion

For the presented games, tests were carried out on different groups of players according to the games' target audience. The educational Tooth defense game was tested on a group of 34 persons of different age in the waiting room of a dental office. Individuals were divided into two groups: for group A, the game recognized emotions and took affective interventions, while for group B no emotions were recognized. Before the beginning and after the game, the participants completed a test examining their knowledge in the field of tooth hygiene. In addition, players answered questions about the level of game satisfaction. The obtained results indicate that the improvement of knowledge was greater for the first group, which may prove the possibility of using affective games in e-learning. Also, a larger percentage of the players from group A rated the game as friendly and less difficult, which indicates the effectiveness of the applied affective interventions. A very interesting fact was that over 82% of the players did not feel any difficulties related to the use of the smartwatch. This feeling would probably be even higher if a smaller smartwatch was used for the younger players. Thus, wearable devices can be successfully used as a HR sensor.

The other two games were tested on a group of 12 students. Opinions expressed by players regarding the affective intervention of the games were very positive. Players particularly appreciated changes in the games' difficulty, when the gameplay became stressful. Reducing the difficulty level allowed to focus on solving the problem giving the players more satisfaction from the gameplay. In particular, changing the pace of moving platforms in Ejcom games was a very positive surprise.

The presented examples of affect-aware games show that using recognition of emotions to control games can be attractive for the players. In particular, this can be used for educational games in order to maintain a player in the flow state to increase the effectiveness of learning. Although the real accuracy of emotion recognition was not measured during the tests, the feedback of the players confirmed the correct direction of affective interventions in most situations. At the same time, players confirmed some disadvantages mentioned earlier connected with the used input channels. This confirms the need for careful selection of the correct input channel depending on the target and predicted use of the scenario platform. On the other hand, it also indicates the possibility of using many different information channels to reduce the disadvantages of using individual ones. In particular, supporting the recognition of emotions based on HR allows for better results in the case of conscious or unconscious control of facial expressions, e.g. in autistic children. Obtained results also lead to the conclusion that the recognition of emotions will play an increasingly important role in the development of computer applications, in particular video games and educational applications.

Acknowledgment. This work was supported by DS Funds of ETI Faculty, Gdansk University of Technology.

References

1. Picard, R.: Affective computing: from laughter to IEEE. IEEE Trans. Affect. Comput. **1**(1), 11–17 (2010)
2. Schell, J.: The Art of Game Design. Morgan Kaufmann Publisher, Elsevier (2008)
3. Gunes, H., Piccardi, M.: Affect recognition from face and body: early fusion vs. late fusion. In: IEEE International Conference on Systems, Man and Cybernetics, vol. 4, pp. 3437–3443 (2003)
4. Szwoch, W.: Using physiological signals for emotion recognition. In: Proceedings of the 6th International Conference on Human System Interaction, pp. 556–561 (2013)
5. Li, L., Chen, J.-H.: Emotion recognition using physiological signals. In: International Conference on Intelligent Information Hiding and Multimedia Signal Processing, pp. 355–358 (2006)
6. Zeng, Z., Pantic, M., Roisman, G., Huang, T.: A survey of affect recognition methods: audio, visual, and spontaneous expressions. IEEE Trans. Pattern Anal. Mach. Intell. **31**(1), 39–58 (2009)
7. Kołakowska, A., Landowska, A., Szwoch, M., Szwoch, W., Wróbel, M.R.: Emotion recognition and its application in software engineering. In: Proceedings of 6th International Conference on Human System Interaction (2013)
8. Landowska, A., Szwoch, M., Szwoch, W.: Methodology of affective intervention design for intelligent systems. Interact. Comput. **28**, 737–759 (2016)
9. Ekman, P., Friesen, W.V., Hager, J.C.: Facial action coding system, A Human Face (2002)
10. Russell, J.A., Mehrabian, A.: Evidence for a three-factor theory of emotions. J. Res. Personal. **11**, 273–294 (1977)

11. Jerritta, S., Murugappan, M., Nagarajan, R., Wan, K.: Physiological signals based human emotion recognition: a review. In: Proceedings of the IEEE 7th International Colloquium on Signal Processing and its Applications (2011)
12. Chen, M., Han, J., Guo, L., Wang, J., Patras, I.: Identifying valence and arousal levels via connectivity between EEG channels. In: International Conference on Affective Computing and Intelligent Interaction (2015)
13. Kołakowska, A.: A review of emotion recognition methods based on keystroke dynamics and mouse movements. In: IEEE Human System Interaction (2013)
14. Maehr, W.: eMotion - Estimation of the User's Emotional State by Mouse Motions, Diploma thesis for Fachhochschule Vorarlberg, Dornbirn, Austria (2005)
15. Szwoch, M.: Design elements of affect aware video games. In: Proceedings of the Mulitimedia, Interaction, Design and Innnovation - MIDI 2015, pp. 1–7 (2015)
16. Witczak, K.: Affect-aware Educational Video Game using Wearable Device. MA thesis, Gdansk University of Technology, Gdansk (2017)
17. Chudy, M., Zyntek, T., Brudek, B.: Affect-aware educational game. MA thesis, Gdansk University of Technology, Gdansk (2017)
18. Karwowski, M., et al.: Affect-aware 3D Platform Game using Unity Engine. MA thesis, Gdansk University of Technology, Gdansk (2018)

Fast Adaptive Binarization with Background Estimation for Non-uniformly Lightened Document Images

Hubert Michalak and Krzysztof Okarma[✉] [ID]

Department of Signal Processing and Multimedia Engineering, Faculty of Electrical Engineering, West Pomeranian University of Technology, Szczecin, 26 Kwietnia 10, 71-126 Szczecin, Poland
okarma@zut.edu.pl

Abstract. Fast and reliable adaptive binarization of unevenly lightened document images is one of the key issues for the Optical Character Recognition (OCR) purposes applied in mobile devices with limited computational power. Considering the document image captured in unknown lighting conditions the use of a single global thresholding in the binarization step makes the text recognition impossible as some parts of it might be lost in the analysed binary image.

On the other hand some well-known adaptive binarization methods e.g. Niblack, Sauvola and their modifications, are computationally demanding and might not be efficiently applied in some applications. Therefore a method for filling the gap between those two approaches is proposed in the paper. It is based on the region based approach utilizing the lighting correction method, in which input data are taken from lighting distribution approximated using reduced resolution images. Obtained binarization results are superior in comparison to typically used adaptive thresholding algorithms in terms of computational speed as well as the final OCR accuracy.

Keywords: Binarization · OCR · Document image analysis

1 Introduction

Fast adaptive binarization of document images acquired in unknown lighting conditions is one of the most relevant steps in many algorithms useful for mobile devices. The most typical example may be the application of the Optical Character Recognition (OCR) for document images acquired by a smartphone camera. However, similar methods may also be useful for the recognition of QR codes or navigation of autonomous mobile robots in varying lighting conditions (e.g. in relatively dark corridors or for line followers using machine vision based navigation).

© Springer Nature Switzerland AG 2019
M. Choraś and R. S. Choraś (Eds.): IP&C 2018, AISC 892, pp. 114–122, 2019.
https://doi.org/10.1007/978-3-030-03658-4_14

For all those applications the first step determining the final results of further processing stages is the binarization allowing to reduce the amount of processed data significantly. As the main computational effort is devoted to the analysis and recognition of shapes or lines on the binary image, the initial thresholding operation should be relatively fast, especially in embedded systems with reduced amount of memory and/or limited computational power.

In some automation and robotic systems the additional illuminators are used to ensure the uniform lighting of the object and simplify the analysis of acquired images. A similar approach for documents is utilised using the 2D flatbed scanners and for such obtained images a single global threshold determined e.g. using popular Otsu's method [14], is often sufficient for high accuracy in text recognition applications. Nevertheless, in uncontrolled lighting conditions more sophisticated adaptive thresholding is necessary to compensate the influence of directional light. Unfortunately, most of such methods require the analysis of the neighbourhood of each pixel to determine the local threshold and therefore their computational demands increase seriously.

The most widely known such adaptive ofby Niblack [12], Sauvola [17], Bradley [1] and Wolf [22]. In the simplest approach proposed by Niblack the local threshold determined for each pixel is dependent on the mean intensity and variance of the small fragment of the image whereas Wolf's method is based on the maximization of the local contrast. Some other modifications have been developed by Gatos [4] and Feng [3].

Alternative attempts to adaptive thresholding based on local calculation of Otsu's threshold have been proposed by Moghaddam [11] and Wen [21]. The former known as AdOtsu requires additional background estimation and analysis of the neighbourhood of each pixel whereas the latter utilizes additional Curvelet transform. A few years later Chou [2] has proposed the use of local Otsu's thresholding with additional use of Support Vector Machines for background regions. Another exemplary computationally demanding algorithm has been proposed by Su [20] where the local adaptive contrast is used with additional edge filtering and post-processing operations. On the other hand, some simplified binarization methods based on the approximated histograms have been proposed utilising the Monte Carlo method [6,7].

Recently, the fast region based approach has been proposed [9,10] for the use with degraded quality documents and non-uniformly lightened images. The local threshold is determined for the 64×64 pixels blocks as 95% of the average intensity lowered by 7 assuming its dynamic range from o to 255. The obtained results have been verified experimentally for widely used DIBCO datasets [15] as well as some unevenly illuminated documents subjected to further Optical Character Recognition.

An overview of state-of-the-art algorithms can be found in some survey papers published by Khurshid [5] and Leedham [8], as well as recently by Samorodova [16], Shrivastava [19] and Saxena [18].

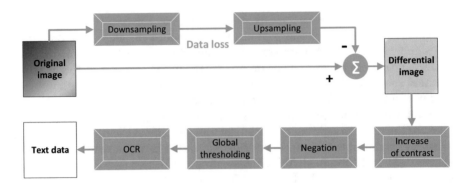

Fig. 1. The workflow of the proposed method

2 Proposed Method and Experimental Verification

Considering the necessity of relatively fast processing caused by computational demands of further text recognition, the binarization method proposed in the paper is based on the region approach allowing to overcome the limitations of pixel based adaptive thresholding.

In the first step of the algorithm the input image is downsampled using one of well known interpolation methods. During the experiments MATLAB's *imresize* function has been used for this purpose with possible application of bilinear or bicubic interpolation and the simple nearest neighbour method. Additionally two versions of Lanczos kernels can be used, based on *sinc* function with the parameter α equal to 2 or 3, as well as box shaped kernel. However, an important parameter is the scaling factor which can be expressed as the size of the window aggregated into a single pixel - in our experiments the square windows have been assumed with the optimal size determined experimentally.

Applying relatively large kernel during downsizing of the image causes the loss of details related to shapes of individual characters. Therefore, after resizing back to the original resolution using the same kernel, the image contains only the low frequency information corresponding to the general distribution of brightness, representing the approximated background.

The next step of the algorithm is the subtraction of such obtained image from the original, followed by increase of its contrast and logical negation. For such obtained image a single global threshold determined e.g. using Otsu's algorithm can be applied. Nevertheless, in our experiments the value of 0.5 has been used to speed up the computations since application of Otsu's method has led to very similar results in terms of the OCR accuracy.

The illustration of the the workflow of the proposed approach is illustrated in Fig. 1 whereas the results of consecutive steps of the algorithm for an exemplary image are presented in Fig. 2.

Since the obtained results strongly depend on the kernel size and type used during resizing, their appropriate choice should be verified experimentally.

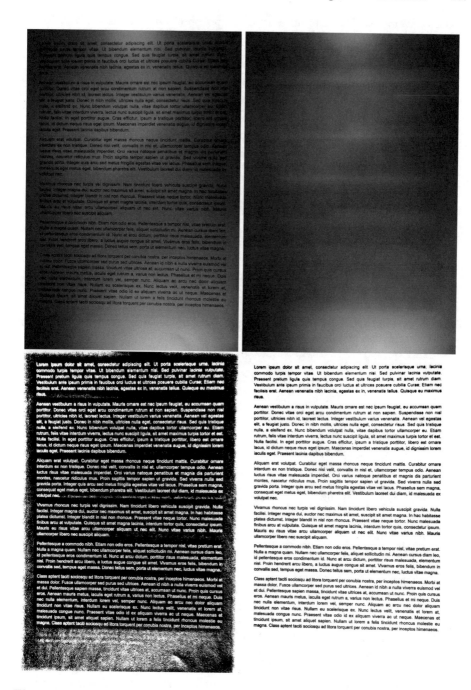

Fig. 2. Results obtained for each stage of the algorithm - from left: input image and approximated background (top), result of subtraction before contrast increasing and final binary image (bottom)

Table 1. Average OCR results and execution times obtained using various binarization methods

Binarization method	F-Measure	Levenshtein distance	Execution time [s]
None	0.4336	2216.06	–
Otsu (global) [14]	0.4514	2240.39	0.0101
Region based [9]	0.9277	46.81	0.0891
Niblack [12]	0.8607	127.94	0.3915
Sauvola [17]	0.9502	32.06	0.3866
Wolf [22]	0.9425	44.28	0.4001
Bradley (mean) [1]	0.9266	61.47	0.1554
Bradley (Gaussian) [1]	0.8620	298.03	1.4655
Proposed (bilinear)	0.9521	31.25	0.0518
Proposed (bicubic)	**0.9571**	**30.36**	0.0579
Proposed (Lanczos2)	**0.9565**	**30.31**	0.0566
Proposed (Lanczos3)	0.9521	32.08	0.0693

Table 2. Average OCR results and execution times obtained for different fonts using various binarization methods

Binarization method	Arial		Times New Roman	
	F-Measure	Levenshtein distance	F-Measure	Levenshtein distance
None	0.4695	2089.44	0.3978	2342.67
Otsu (global) [14]	0.4598	2131.78	0.4420	2349.00
Region based [9]	0.9218	41.94	0.9337	51.67
Niblack [12]	0.8560	111.17	0.8653	144.72
Sauvola [17]	0.9476	28.72	0.9527	35.39
Wolf [22]	0.9398	30.72	0.9451	57.83
Bradley (mean) [1]	0.9204	68.11	0.9328	54.83
Bradley (Gaussian) [1]	0.8684	284.50	0.8555	311.56
Proposed (bilinear)	0.9613	24.17	0.9428	38.33
Proposed (bicubic)	0.9641	22.28	0.9501	38.44
Proposed (Lanczos2)	0.9640	22.00	0.9489	38.61
Proposed (Lanczos3)	0.9575	24.17	0.9468	40.00

Therefore, a dataset consisting of 36 differently illuminated document images has been created. All of them have been captured by a standalone camera in various lighting conditions (i.e. in the presence of some directional light sources located somewhere over the document). Documents contain a version of well known "Lorem ipsum" text generated using a dedicated website,[1] printed using

---

[1] pl.lipsum.com.

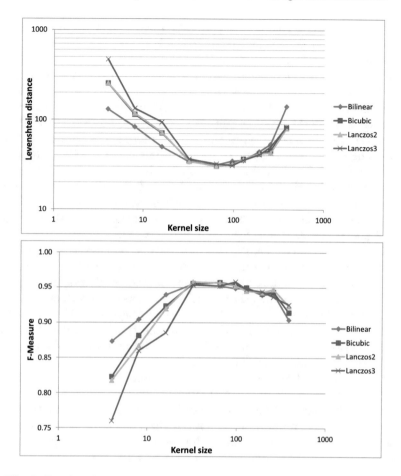

Fig. 3. Results obtained for various kernels used during preprocessing step

two popular fonts (Arial and Times New Roman), using their bold versions as well, and therefore the correct result of text recognition is known.

The verification of the binarization method's influence on the OCR accuracy has been conducted by the computation of Levenshtein distance between the recognized text and the "ground truth" as well as calculating the F-Measure considering the number of correctly and incorrectly recognized characters. Levenshtein distance refers to the minimum number of text edits (insertions, deleting or replacements of individual characters) required to convert the recognized text into the "ground truth". Supplementary, the F-Measure, known also as F1-score, is defined as:

$$\text{FM} = 2 \cdot \frac{PR \cdot RC}{PR + RC} \cdot 100\%, \tag{1}$$

where Precision (PR) is computed as the ratio of true positives to the sum of all positives, and Recall (RC) is defined as the ratio of true positives to the sum of true positives and false negatives.

Typically, for the binarization quality assessment, this metric is applied for single pixels of the binary images [13], however it is also often considered as general classification quality metric, also for text recognition purposes.

To achieve the best possible results both these metrics have been calculated assuming the use of different types and size of interpolation kernels during the preprocessing step, preceding binarization and launching the OCR software. In all experiments the free Tesseract OCR engine has been used which had been invented initially by Hewlett Packard and University of Nevada and further developed by Google. As shown in Fig. 3, the best results have been achieved for Lanczos and bicubic 64×64 pixels kernels (Lanczos2 denotes the kernel obtained for $\alpha = 2$ and Lanczos3 stands for $\alpha = 3$ respectively).

Having optimized the kernel size, we have compared the OCR results obtained using the proposed binarization method with the use of some other popular adaptive thresholding algorithms. To illustrate additionally the advantages of the proposed approach the execution time has also been compared using the same MATLAB environment. The OCR quality measures (Levenshtein distance and F-Measure) together with the running time of various binarization algorithms averaged for all 36 images are presented in Table 1, whereas Table 2 shows the results obtained for different fonts.

As can be easily observed, the proposed approach is much faster than the popular adaptive methods and comparable to recently proposed region based approach, leading also to better OCR results. Comparing the results obtained for two popular fonts it can also be noticed that, similarly as for most of the other algorithms, better recognition results are achieved for Arial font containing simplified font shapes in comparison to Times New Roman.

3   Conclusions and Future Work

The proposed fast method of document image binarization can be efficiently applied for text recognition purposes from non-uniformly illuminated document images. Presented approach combines short execution time with good OCR accuracy obtained after binarization and outperforms popular adaptive thresholding algorithms as well as recently proposed region based method. Therefore one of its potential areas of applications may be related to preprocessing of document images captured by smartphone cameras.

Our future work will concentrate on the verification of the proposed approach for various font shapes, together with bold and italic versions, as well as further modifications of the fast background estimation.

References

1. Bradley, D., Roth, G.: Adaptive thresholding using the integral image. J. Gr. Tools **12**(2), 13–21 (2007)
2. Chou, C.H., Lin, W.H., Chang, F.: A binarization method with learning-built rules for document images produced by cameras. Pattern Recognit. **43**(4), 1518–1530 (2010)
3. Feng, M.L., Tan, Y.P.: Adaptive binarization method for document image analysis. In: Proceedings of the 2004 IEEE International Conference on Multimedia and Expo (ICME), vol. 1, pp. 339–342 (2004)
4. Gatos, B., Pratikakis, I., Perantonis, S.: Adaptive degraded document image binarization. Pattern Recognit. **39**(3), 317–327 (2006)
5. Khurshid, K., Siddiqi, I., Faure, C., Vincent, N.: Comparison of Niblack inspired binarization methods for ancient documents. In: Document Recognition and Retrieval XVI, vol. 7247, pp. 7247–7247–9 (2009)
6. Lech, P., Okarma, K.: Fast histogram based image binarization using the Monte Carlo threshold estimation. In: Chmielewski, L.J., Kozera, R., Shin, B.S., Wojciechowski, K. (eds.) Computer Vision and Graphics. LNCS, vol. 8671, pp. 382–390. Springer International Publishing, Switzerland (2014)
7. Lech, P., Okarma, K.: Optimization of the fast image binarization method based on the Monte Carlo approach. Elektron. Ir Elektrotech. **20**(4), 63–66 (2014)
8. Leedham, G., Yan, C., Takru, K., Tan, J.H.N., Mian, L.: Comparison of some thresholding algorithms for text/background segmentation in difficult document images. In: Proceedings of the 7th International Conference on Document Analysis and Recognition, ICDAR 2003, pp. 859–864 (2003)
9. Michalak, H., Okarma, K.: Region based adaptive binarization for optical character recognition purposes. In: 2018 International Interdisciplinary PhD Workshop (IIPhDW), pp. 361–366 (2018)
10. Michalak, H., Okarma, K.: Fast adaptive image binarization using the region based approach. In: Silhavy, R. (ed.) Artificial Intelligence and Algorithms in Intelligent Systems, AISC, vol. 764, pp. 79–90. Springer International Publishing (2019)
11. Moghaddam, R.F., Cheriet, M.: AdOtsu: an adaptive and parameterless generalization of Otsu's method for document image binarization. Pattern Recognit. **45**(6), 2419–2431 (2012)
12. Niblack, W.: An Introduction to Digital Image Processing. Prentice Hall, Englewood Cliffs (1986)
13. Ntirogiannis, K., Gatos, B., Pratikakis, I.: Performance evaluation methodology for historical document image binarization. IEEE Trans. Image Process. **22**(2), 595–609 (2013)
14. Otsu, N.: A threshold selection method from gray-level histograms. IEEE Trans. Syst. Man Cybern. **9**(1), 62–66 (1979)
15. Pratikakis, I., Zagoris, K., Barlas, G., Gatos, B.: ICDAR 2017 Document Image Binarization COmpetition (DIBCO 2017) (2017). https://vc.ee.duth.gr/dibco2017/
16. Samorodova, O.A., Samorodov, A.V.: Fast implementation of the Niblack binarization algorithm for microscope image segmentation. Pattern Recognit. Image Anal. **26**(3), 548–551 (2016)
17. Sauvola, J., Pietikäinen, M.: Adaptive document image binarization. Pattern Recognit. **33**(2), 225–236 (2000)

18. Saxena, L.P.: Niblack's binarization method and its modifications to real-time applications: a review. Artif. Intell. Rev. 1–33 (2017)
19. Shrivastava, A., Srivastava, D.K.: A review on pixel-based binarization of gray images. In: AISC, vol. 439, pp. 357–364. Springer, Singapore (2016)
20. Su, B., Lu, S., Tan, C.L.: Robust document image binarization technique for degraded document images. IEEE Trans. Image Process. 22(4), 1408–1417 (2013)
21. Wen, J., Li, S., Sun, J.: A new binarization method for non-uniform illuminated document images. Pattern Recognit. 46(6), 1670–1690 (2013)
22. Wolf, C., Jolion, J.M.: Extraction and recognition of artificial text in multimedia documents. Form. Pattern Anal. Appl. 6(4), 309–326 (2004)

Automatic Colour Independent Quality Evaluation of 3D Printed Flat Surfaces Based on CLAHE and Hough Transform

Jarosław Fastowicz and Krzysztof Okarma$^{(\boxtimes)}$ (iD)

Department of Signal Processing and Multimedia Engineering, Faculty of Electrical Engineering, West Pomeranian University of Technology, Szczecin, 26 Kwietnia 10, 71-126 Szczecin, Poland
okarma@zut.edu.pl

Abstract. In this paper a novel approach to quality evaluation of flat surfaces of the 3D printed objects has been discussed. The proposed approach utilizes the regularity of the layers visible on the printed surfaces which are extracted using Hough transform. Nevertheless, due to the variety of filament's colours, the application of a single metric which can be automatically computed, being equivalent to perceived surface quality, requires the additional preprocessing operations. For this purpose Contrast Limited Adaptive Histogram Equalization (CLAHE) has been used together with additional compensation of the metric for bright filaments. Achieved results for the database of 88 scanned samples are encouraging and allow a reliable quality assessment of 3D surfaces for various filaments.

Keywords: 3D prints · Quality assessment · Hough transform
CLAHE

1 Introduction

Additive manufacturing becomes one of the most significant technologies changing the production and business related to the dynamic growth of "Industry 4.0". Increasing popularity and availability of the 3D printers, especially based on fused deposition modelling (FDM) technology, which can be considered also as devices for everyday home use, as well as various types and colours of filaments, causes also some new scientific and technological challenges. Apart from development of new materials used as filaments, some recent areas of research in this field are related to cybersecurity and safety [13,15,16], process monitoring [3,7], fault detection [1,2,4,17] or finding some of the microdefects [14].

Some of the above mentioned issues are partially solved using the machine vision algorithms allowing the observation and automatic analysis of the 3D printing process. Nevertheless, usually their application is limited to the emergency stop in the case of fault detection without the on-line quality monitoring of the printed object's quality. Development of a reliable automatic quality

© Springer Nature Switzerland AG 2019
M. Choraś and R. S. Choraś (Eds.): IP&C 2018, AISC 892, pp. 123–131, 2019.
https://doi.org/10.1007/978-3-030-03658-4_15

assessment of the 3D printed surfaces and its further on-line application for the 3D printers would allow to make some possible corrections during the printing process or stopping the printing process for low quality samples. Considering the long time (usually hours rather than minutes) necessary for 3D printing of even relatively small objects in various applications (e.g. biomedical [18]), such an approach would allow saving both time and material.

Nevertheless, the first step should be the development of a reliable metric which could be applied and its verification which can also be made off-line. Some attempts presented recently utilize texture analysis [5], also based on the Gray-Level Co-occurrence Matrix (GLCM) [8], similarity based image quality assessment (IQA) methods [9,11] as well as image entropy [6,10]. However, most of the methods should be additionally tuned for various colours of filaments as the obtained results are different for different materials. Additionally, some of them, e.g. based on GLCM, have relatively high computational demands and cannot be efficiently applied in real-time.

Currently the best results have been obtained using the entropy based approach [10] providing colour independent metric, although tested using a smaller dataset consisting of 18 PLA samples with 5 colours. Therefore, the goal of this paper is to propose a new colour independent quality metric and verify its properties for a larger database of scanned 3D printed surfaces.

2   Proposed Method

Considering the principle of operation of low budget FDM 3D printers, the high quality sample should contain surfaces with visible regular linear patterns. For some types of materials (mainly ABS) their further possible removal requires additional operations, however during the printing or directly after manufacturing the characteristic stripes are well visible. Assuming the use of a camera observing the side surface of the 3D printed object and the flatness of this surface, the visible horizontal lines should be straight and their detection on the image should be possible using the popular Hough transform.

Since the detection of long straight lines for the whole sample might be troublesome, much better results can be obtained by the analysis of some chosen fragments of an image. The influence of the small rotations of the scanned sample (or cameras for on-line applications) as well as some relatively small imperfections or local lighting changes may influence the results of Hough transform significantly as it has been verified experimentally. Therefore, it has been decided to apply the Monte Carlo method [12] for random draw of the positions of some fragments of the sample subjected to further analysis. Choosing the reasonable number of drawn regions and their size, the computational demands of the algorithm may also be additionally decreased. During the conducted experiments, the best fully repeatable results has been obtained using 40 regions of 300×150 pixels where the resolution of full images in the database is 1662×1662 pixels so the number of pixels being analysed has been reduced initially by about one third. The illustration of such regions and detected lines for one of low quality

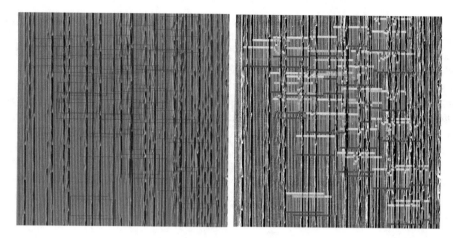

Fig. 1. An exemplary rotated low quality sample with marked regions randomly chosen by the Monte Carlo method (left) and the results of binarization and line detection using Hough transform (right)

samples is presented in Fig. 1 (all samples has been rotated by 90° after scanning). As can be observed the number of detected lines would be higher for better quality samples. Therefore, the initial quality metric has been based on the average length of detected lines in all regions.

Nevertheless, for various colours of samples converted to grayscale according to popular ITU Recommendation BT.601-7 (used in MATLAB's *rgb2gray* function), results differ significantly, mainly due to differences in contrast, and therefore the additional preprocessing is necessary. For this purpose the application of Contrast Limited Adaptive Histogram Equalization (CLAHE) algorithm [19] has been examined.

The choice of the valid parameters of the CLAHE algorithm has been made as the next step of the experiments. The best discrimination between the high and low quality samples has been obtained using 256 bins with assumed exponential histogram distribution with the parameter $\alpha = 0.9$ and clip limit set to 0.1. The number of tiles into which the image has been divided has been verified experimentally as well. Considering the resolution of input images and the high number of visible layers, the default values from the MATLAB's *adapthisteq* function equal to 8×8 tiles can be treated as appropriate for the analysed samples.

Since the results of further line detection are also dependent on the binarization algorithm, some popular thresholding methods have been tested as well. To prevent the use of sophisticated algorithms for on-line monitoring purposes, we have focused on the global thresholding and therefore well-known Otsu's binarization seems to be enough for images previously subjected to CLAHE algorithm.

Regardless of the advantages of the proposed preprocessing approach, the average length of the detected lines for the brightest samples is noticeably smaller and therefore the additional correction coefficient has been proposed for the samples with the normalized luminance $Y > 0.8$ calculated according to the mentioned above ITU BT.601-7 recommendation. The value of the coefficient added to the computed average length of lines determined experimentally is equal to $0.02 \cdot M \cdot Y$ assuming the normalized luminance Y within the range $\langle 0; 1 \rangle$ and M being the maximum length of the line dependent on the size of the region ($M = 150$ in our experiments). However, for high quality bright samples the overall result may exceed the value of M. To allow more convenient interpretation of results it is recommended to use the natural logarithm of such calculated value - for the assumed size of the regions the threshold value between high and low quality samples is then set to 5.

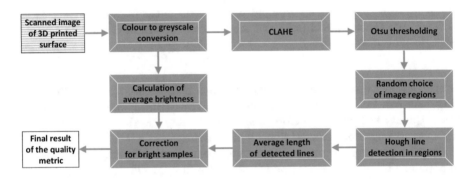

Fig. 2. The workflow of the proposed method

The simplified workflow of the proposed method is illustrated in Fig. 2.

3   Experiments and Discussion of Results

To verify the possibilities of application of various image analysis methods for this purpose a dedicated database of 88 flat samples has been built, containing some 3D prints with intentionally introduced distortions during printing. It has been prepared using two fused deposition modelling printers and several colours of plastic filaments of two popular types - PLA (Polylactic Acid) and ABS (Acrylonitrile Butadiene Styrene). Considering different physical properties of both materials, the visible distortions have been forced by the changes of the temperature and filament's delivery speed during manufacturing. Some of the ABS samples contain some small cracks and all the obtained surfaces have been assessed as high, moderately high, moderately low or low quality. For the off-line analysis to prevent the influence of unknown lighting conditions, all the 3D printed objects have been scanned using a flatbed scanner using the density of 1200 dpi.

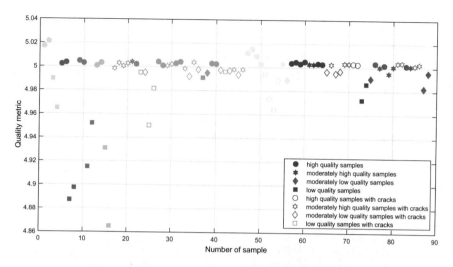

Fig. 3. Results obtained for the 88 analysed samples using the proposed approach

Fig. 4. Exemplary ABS moderate high quality samples no. 17 (left) and 18 (right)

The results obtained using the proposed method are shown in Fig. 3 where the quality metric is expressed as the natural logarithm of the average length of detected lines with additional correction for bright samples. As it can be noticed for high quality white and yellow samples, due to the use of additional correction coefficient, the obtained values may exceed the upper limit of ln(M).

The colours in the plot shown in Fig. 3 represent the colours of individual samples (however the first 4 have been slightly darkened for better visibility). The first 16 samples have been obtained for the PLA filaments (4 colours – two images of high and two of low quality samples, technically being both sides of the same 3D prints). The rest of the samples have been obtained using the ABS filaments where empty symbols denote the presence of cracks. The shapes of the symbols represent the subjective quality of the surface - circles for the high quality, stars - moderately high, diamonds - moderately low and squares for the low quality.

Analysing the obtained values, the satisfactory classification of most of the samples can be noticed, especially for high and low quality samples. Nevertheless, the results obtained for some of the samples marked as moderately high quality surfaces should be briefly discussed. Considering for example the first two brown samples (no. 17 and 18), being in fact the two sides of the same 3D print presented in Fig. 4, marked as moderately high quality with cracks, it can be observed that the left image contains more contaminations leading to slightly lower value of the quality metric (4.9984 vs. 5.0029). Nevertheless, both of those sides have been subjectively assessed as similar using the assumed four-level scale.

Fig. 5. Exemplary ABS moderate high quality samples no. 33 (upper left), 44 (upper right), 46 (lower left) and 79 (lower right)

Some other interesting examples are the samples no. 33 (quality 4.9985), 44 (quality 4.9980), 46 (quality 4.9980) and 79 (quality 4.9951) presented in Fig. 5. The last of them seems to have a significant higher visual quality, although due to the presence of many small contaminations a proper detection of some of the lines representing consecutive layers might be troublesome and therefore the overall result is lowered. Therefore in some cases the obtained results might be a bit surprising. Nevertheless, such sensitivity to even small contaminations may be considered as one of the advantages of the proposed approach being the first step towards the quality metric linearly related to the quantity of visible distortions.

4 Conclusions and Future Work

The method of automatic colour independent quality assessment of 3D printed surfaces provides encouraging results, being well correlated with subjective quality perception. Its main advantage is the independence on colour and type of popular filaments used for low budget 3D printing, however in some cases the obtained results may vary, mainly in the presence of small contaminations for moderate quality samples. It is partially caused by the use of the Monte Carlo method [12] randomly limiting the analysed area of the sample.

The planned future development will be concentrated on further tuning of the method e.g. by a gentle image pre-filtering allowing to remove some very small, almost invisible contaminations, which in fact do not influence the properties of the final product. Such a filtration would not influence the cracks and more serious distortions allowing to obtain even more reliable results e.g. for images similar to the sample no. 79 discussed above.

Another direction of our future research will be the combination of the presented method with some other recently investigated approaches, e.g. based on texture analysis [5] and entropy [10] as well as some experiments related to the further decrease of the amount of necessary computations.

References

1. Chauhan, V., Surgenor, B.: A comparative study of machine vision based methods for fault detection in an automated assembly machine. Procedia Manuf. **1**, 416–428 (2015)
2. Chauhan, V., Surgenor, B.: Fault detection and classification in automated assembly machines using machine vision. Int. J. Adv. Manuf. Technol. **90**(9), 2491–2512 (2017)
3. Cheng, Y., Jafari, M.A.: Vision-based online process control in manufacturing applications. IEEE Trans. Autom. Sci. Eng. **5**(1), 140–153 (2008)
4. Fang, T., Jafari, M.A., Bakhadyrov, I., Safari, A., Danforth, S., Langrana, N.: Online defect detection in layered manufacturing using process signature. In: Proceedings of IEEE International Conference on Systems, Man and Cybernetics, San Diego, California, USA, vol. 5, pp. 4373–4378 (1998)
5. Fastowicz, J., Okarma, K.: Texture based quality assessment of 3D prints for different lighting conditions. In: Chmielewski, L.J., Datta, A., Kozera, R., Wojciechowski, K. (eds.) Computer Vision and Graphics. ICCVG 2016. LNCS, vol. 9972, pp. 17–28. Springer International Publishing (2016)

6. Fastowicz, J., Okarma, K.: Entropy based surface quality assessment of 3D prints. In: Silhavy, R., Senkerik, R., Kominkova Oplatkova, Z., Prokopova, Z., Silhavy, P. (eds.) Artificial Intelligence Trends in Intelligent Systems. CSOC 2017. AISC, vol. 573, pp. 404–413. Springer International Publishing (2017)

7. Gardner, M.R., Lewis, A., Park, J., McElroy, A.B., Estrada, A.D., Fish, S., Beaman, J.J., Milner, T.E.: In situ process monitoring in selective laser sintering using optical coherence tomography. Opt. Eng. **57**, 57-1–57-5 (2018)

8. Okarma, K., Fastowicz, J.: No-reference quality assessment of 3D prints based on the GLCM analysis. In: Proceedings of the 2016 21st International Conference on Methods and Models in Automation and Robotics (MMAR), pp. 788–793 (2016)

9. Okarma, K., Fastowicz, J.: Quality assessment of 3D prints based on feature similarity metrics. In: Choraś, R.S. (ed.) Image Processing and Communications Challenges 8. IP&C 2016. AISC, vol. 525, pp. 104–111. Springer International Publishing (2017)

10. Okarma, K., Fastowicz, J.: Color independent quality assessment of 3D printed surfaces based on image entropy. In: Kurzynski, M., Wozniak, M., Burduk, R. (eds.) Proceedings of the 10th International Conference on Computer Recognition Systems CORES 2017. AISC, vol. 578, pp. 308–315. Springer International Publishing (2018)

11. Okarma, K., Fastowicz, J., Tecław, M.: Application of structural similarity based metrics for quality assessment of 3D prints. In: Chmielewski, L.J., Datta, A., Kozera, R., Wojciechowski, K. (eds.) Computer Vision and Graphics. ICCVG 2016. LNCS, vol. 9972, pp. 244–252. Springer International Publishing (2016)

12. Okarma, K., Lech, P.: Monte Carlo based algorithm for fast preliminary video analysis. In: Bubak, M., van Albada, G.D., Dongarra, J., Sloot, P.M.A. (eds.) Computational Science - ICCS 2008. LNCS, vol. 5101, pp. 790–799. Springer, Heidelberg (2008)

13. Straub, J.: 3D printing cybersecurity: detecting and preventing attacks that seek to weaken a printed object by changing fill level. In: Proceedings of SPIE – Dimensional Optical Metrology and Inspection for Practical Applications VI, vol. 10220, Anaheim, CA, USA, pp. 102200O-1–102200O-15 (2017)

14. Straub, J.: An approach to detecting deliberately introduced defects and micro-defects in 3D printed objects. In: Proceedings of SPIE – Pattern Recognition and Tracking XXVII, Anaheim, CA, USA, vol. 10203, pp. 102030L-1–102030L-14 (2017)

15. Straub, J.: Identifying positioning-based attacks against 3D printed objects and the 3D printing process. In: Proceedings of SPIE – Pattern Recognition and Tracking XXVII, Anaheim, CA, USA, vol. 10203, pp. 1020304-1–1020304-13 (2017)

16. Straub, J.: Physical security and cyber security issues and human error prevention for 3D printed objects: detecting the use of an incorrect printing material. In: Proceedings of SPIE – Dimensional Optical Metrology and Inspection for Practical Applications VI, Anaheim, CA, USA, vol. 10220, pp. 102200K-1–102200K-16 (2017)

17. Szkilnyk, G., Hughes, K., Surgenor, B.: Vision based fault detection of automated assembly equipment. In: Proceedings of the ASME/IEEE International Conference on Mechatronic and Embedded Systems and Applications, Parts A and B, Washington, DC, USA, vol. 3, pp. 691–697 (2011)

18. Ware, H.O.T., Farsheed, A.C., Baker, E., Ameer, G., Sun, C.: Fabrication speed optimization for high-resolution 3D-printing of bioresorbable vascular scaffolds. In: Procedia CIRP 65 (3rd CIRP Conference on BioManufacturing), pp. 131–138 (2017)
19. Zuiderveld, K.: Contrast limited adaptive histogram equalization. In: Heckbert, P.S. (ed.) Graphics Gems IV, pp. 474–485. Academic Press Professional Inc. (1994)

Using Segmentation Priors to Improve the Video Surveillance Person Re-Identification Accuracy

Dominik Pieczyński, Marek Kraft$^{(\boxtimes)}$, and Michał Fularz

Institute of Control, Robotics and Information Engineering,
Poznań University of Technology, Piotrowo 3A, 60-965 Poznań, Poland
marek.kraft@put.poznan.pl

Abstract. In this paper, a method for improving the quality of person re-identification results is presented. The method is based on the assumption, that including segmentation information into re-identification pipeline suppresses the influence of the background and discards the automated detections that are of poor quality due to occlusions, misplaced regions of interest (ROI), multiple persons found within a single ROI etc. Assuming that a joint detector-segmented approach is used, the additional cost associated with the use of the proposed approach is very low.

1 Introduction

Person re-identification is one of the most prominent tasks in video surveillance systems. First introduced as a computer vision research problem in [13] in the context of human-robot interaction, it was defined as a task to 're-identify a person when it leaves the field of view and re-enters later'. Since then, it has found its way into video surveillance systems and became popular in the community due to its application and research significance. Pioneering approaches were usually based on colour and texture information. Fusion of information from multiple views followed soon after [2], along with the approaches that aimed at decreasing the influence of background by performing some kind of segmentation [4]. Following the recent research trends in computer vision, deep learning based approaches first appeared in [11]. Introduction of those new approaches caused a breakthrough change in re-identification accuracy, reaching over 80% rank-1 accuracy on challenging datasets with over 1000 individuals registered across multiple views [7,8,16].

That being said, person re-identification is still a challenging problem. Datasets used to train and test the deep learning approaches are based on automatic detection, which might give rise to problems shown in Fig. 1. Although the presented example image pairs are assigned to the same individuals in the dataset, getting correct re-identification using these image pairs would certainly be less likely. On the other hand, a video surveillance system operating under realistic conditions must also cope with these issues.

M. Choraś and R. S. Choraś (Eds.): IP&C 2018, AISC 892, pp. 132–139, 2019.
https://doi.org/10.1007/978-3-030-03658-4_16

Fig. 1. Example issues associated with datasets created using automated sample collection – from the left: incomplete person within the region of interest (ROI), misplaced ROI containing a significant portion of background, severe image quality degradation due to blur, occlusion by another person. Images taken from the MARS dataset [14]

In this paper, the influence of using segmentation priors on the accuracy of person re-identification is evaluated. The rationale behind the idea is the fact, that in video surveillance systems re-identification is usually preceded by object detection. Since methods for joint object detection and segmentation are available [5], the detection bounding box and its internal segmentation result can be used as an input to a simple rule-based system, that rules out problematic cases so that the re-identification procedure is not executed for potentially invalid image pairs.

2   Proposed Approach

The proposed approach is based on two key concepts: the re-identification neural network and prior detection and segmentation operation based on the Mask-RCNN method.

The re-identification neural network model follows the straightforward siamese convolutional neural network architecture, using ResNet-50 [6] backend for both feature detection paths. The generalised structure of the network is shown in Fig. 2. Such simple architecture was selected on purpose to eliminate the influence of other techniques used in re-identification systems such as re-ranking [17], metric learning [9] or multiple query [15].

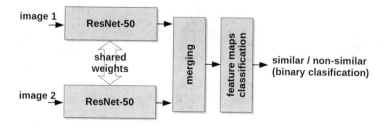

Fig. 2. Block diagram of re-identification neural network architecture

134   D. Pieczyński et al.

Adam optimiser with Nesterov momentum [3] was used during the training. All hyperparameters were set to default values except for the learning rate of $3e - 04$. Binary cross-entropy was selected as a loss function. The data was randomly split into training and validation sets so that images of 181 persons were used for training and images of 33 persons were used for validation. To benefit from transfer learning to speed up the training and improve the results [12], we used the ResNet-50 variant that was pre-trained on the ImageNet for initialisation. The network was trained for 300 epochs with early stopping if no improvement of validation loss was noted in the last 50 epochs. The model's weights were automatically saved when there was an improvement either in validation loss or in validation accuracy. Both the segmented and non-segmented versions of images were used for training and testing to compare the accuracy of both approaches.

Incorporating segmentation directly into the object of interest detection has several advantages in the context of person re-identification. First, the re-identification can be constrained to the object of interest, eliminating the influence of background that can change significantly across multiple views of an object. Second, the coefficients such as ROI fill ratio can be used to rule out the imperfect images from processing.

Amongst the novel methods for object detection, Mask-RCNN [5] is especially interesting. It is able to detect multiple entities and perform automated segmentation inside their bounding boxes. This means that using this approach as a detector or an automated method for generating new re-identification dataset extends the standard bounding boxes data with masks that allow testing of a background suppression effectiveness (see Fig. 3).

Fig. 3. Mask-RCNN, aside from performing object detection, effectively differentiates between objects and the background, even if the image quality is imperfect

Mask-RCNN is an extension to Faster-RCNN [10] network. It uses the same principles to detect objects, but adds additional, parallel branch that performs segmentation.

The idea behind Faster-RCNN is to detect objects in two stages. The network performs its operations using convolutional feature maps as the input. An important characteristic of such approach is that those maps can be generated by a variety of network models. This makes Faster-RCNN decoupled from network's base and allows using both simpler and more complex models depending on the need and available hardware.

The first stage called Region Proposal Network (RPN) finds predefined number of rectangular regions that may contain objects. In order to perform this tasks the RPN uses anchors, which are fixed sized bounding boxes that are moved over the image using configurable stride (Figs. 4 and 5).

Fig. 4. Faster-RCNN anchors. In this example the network is configured to use 9 anchors (3 bounding boxes in 3 scales denoted by different line styles)

As the anchors number, aspect ratios and scales can be adjusted prior to the network's training, this method provides a versatile way to select multiple objects of different shapes.

The second stage of Faster-RCNN, which shares network's weights with the first stage, performs feature extraction for every proposed candidate and proceeds with classification as an object or a background and bounding-box regression, which adjusts initial anchor's coordinates.

Mask-RCNN extends the second stage of Faster-RCNN with a parallel, fully convolutional branch that outputs a binary mask for each region of interest. This way the computational demand does not increase dramatically and the network retains Faster-RCNN properties. Most importantly, the *head*, performing bounding-box recognition and mask prediction, is still decoupled from the *backbone*, used for obtaining a feature map.

Mask-RCNN can therefore be used as both the pedestrian detector and the segmenter in re-identification systems. This method can also be extended with

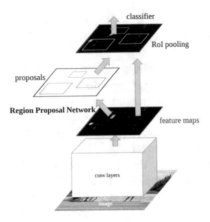

Fig. 5. Faster-RCNN general architecture. Image source: [10]

tracking capabilities (retaining person label between video frames). Example results of applying Mask-RCNN in a typical video surveillance setting are shown in Fig. 6.

Fig. 6. Example bounding boxes with segmentation results generated fully automatically by Mask-RCNN

The following criteria were used to discard candidate images based on the detection and segmentation result within the ROI:

- no object of the 'person' class is detected within the ROI, which may indicate misplaced ROI,
- more than one object of the 'person' class is detected (with segmentation) within the ROI, which might indicate occlusion,

- area of segmented foreground (pixels belonging to the 'person' class) within the ROI is below 14% of the ROI area, which might indicate misplaced ROI with a significant portion of background or occlusion by environment elements,
- the lengths of the sides of the ROI do not the condition: $1.8 < height/width < 2.2$, which should indicate that a complete person in upright position is not present within the ROI.

3   Evaluation Results

The MARS image dataset was used for evaluation [14]. The dataset contains automatically collected data for 1261 different individuals. Each individual is observed by up to 6 different cameras, although not all persons are visible in all the cameras. Altogether, the dataset contains 1 191 003 images.

For the purpose of this research, a subset of MARS dataset was chosen. Only persons visible by 4 or more different cameras were used. This makes the overall re-identification task more difficult, but size reduction allows for faster training and testing of the method. Overall 214 (16%) individuals with 572 647 (48%) images were used.

The ResNet-50 network was trained in few different scenarios, that is:

- on a whole reduced dataset, to get the baseline performance of this network,
- on a reduced dataset, where problematic cases were filtered out based on the rules described previously,
- on a reduced dataset, where problematic cases were filtered out based on the rules described previously and the images were segmented.

Fig. 7. Example images automatically removed from test set. Top row detections in the left did not meet aspect ratio criterion, top row detections in the right presented multiple persons and the images in the bottom row, shown as pairs of original image and segmented one, had insufficient fill ratio

Images were filtered using an implementation of Mask-RCNN [1]. Examples are shown in Fig. 7. The applied rules resulted in removal of large portion of the dataset - only 37% of train and 38% of test images were left in the pool. It should be noted, that while evidently wrong images were removed, a lot of fine looking ones were falsely discarded.

Results obtained are shown in the Table 1. Given the query image, the predictions were performed and the similarity was scored and sorted for all target persons in the database. Rank n accuracy means that the classification was marked successful if the correct person was present in n most similar persons chosen by the network.

Table 1. ResNet-50-based model rank accuracy for different types of data preprocessing

Data preprocessing	$r-1$	$r-5$	$r-10$	$r-20$
None	0.534	0.823	0.901	0.943
Filtration	0.593	0.864	0.922	0.957
Filtration + Segmentation	**0.635**	**0.882**	**0.932**	**0.962**

Each of the proposed steps resulted in improvement in accuracy. Rank-1 increased by 5.94% points as a result of applying the filtering and by further 4.12% point by addition of segmentation. Overall, the final solution that employs both processing steps achieved 10.06% point (relative increase of 19%) improvement compared to baseline solution.

4   Conclusions

A method for improving the accuracy of person re-identification was proposed. The method uses segmentation priors to filter out the problematic image pairs, whose analysis might give rise to errors. The method is based on a variety of simple characteristics, whose computation is possible under the assumption that the joint detection and segmentation approach is used as a prior processing step. The computational cost of the method is low, yet the improvement in re-identification accuracy is significant. Moreover, the method can be used in conjunction with a wide range of existing re-identification approaches.

References

1. Abdulla, W.: Mask R-CNN for object detection and instance segmentation on Keras and TensorFlow (2017). https://github.com/matterport/Mask_RCNN
2. Bazzani, L., Cristani, M., Perina, A., Farenzena, M., Murino, V.: Multiple-shot person re-identification by HPE signature. In: 2010 20th International Conference on Pattern Recognition (ICPR), pp. 1413–1416. IEEE (2010)
3. Dozat, T.: Incorporating Nesterov momentum into Adam (2015)

4. Farenzena, M., Bazzani, L., Perina, A., Murino, V., Cristani, M.: Person re-identification by symmetry-driven accumulation of local features. In: 2010 IEEE Conference on Computer Vision and Pattern Recognition (CVPR), pp. 2360–2367. IEEE (2010)
5. He, K., Gkioxari, G., Dollár, P., Girshick, R.: Mask R-CNN. In: 2017 IEEE International Conference on Computer Vision (ICCV), pp. 2980–2988. IEEE (2017)
6. He, K., Zhang, X., Ren, S., Sun, J.: Deep residual learning for image recognition. arXiv preprint arXiv:1512.03385 (2015)
7. Hermans, A., Beyer, L., Leibe, B.: In defense of the triplet loss for person re-identification. arXiv preprint arXiv:1703.07737 (2017)
8. Li, W., Zhu, X., Gong, S.: Person re-identification by deep joint learning of multi-loss classification. arXiv preprint arXiv:1705.04724 (2017)
9. Liao, S., Hu, Y., Zhu, X., Li, S.Z.: Person re-identification by local maximal occurrence representation and metric learning. In: Proceedings of the IEEE Conference on Computer Vision and Pattern Recognition, pp. 2197–2206 (2015)
10. Ren, S., He, K., Girshick, R., Sun, J.: Faster R-CNN: towards real-time object detection with region proposal networks. IEEE Trans. Pattern Anal. Mach. Intell. **6**, 1137–1149 (2017)
11. Yi, D., Lei, Z., Liao, S., Li, S.Z.: Deep metric learning for person re-identification. In: 2014 22nd International Conference on Pattern Recognition (ICPR), pp. 34–39. IEEE (2014)
12. Yosinski, J., Clune, J., Bengio, Y., Lipson, H.: How transferable are features in deep neural networks? In: Advances in Neural Information Processing Systems, pp. 3320–3328 (2014)
13. Zajdel, W., Zivkovic, Z., Krose, B.: Keeping track of humans: have I seen this person before? In: Proceedings of the 2005 IEEE International Conference on Robotics and Automation 2005, ICRA 2005, pp. 2081–2086. IEEE (2005)
14. Zheng, L., Bie, Z., Sun, Y., Wang, J., Su, C., Wang, S., Tian, Q.: MARS: a video benchmark for large-scale person re-identification. In: European Conference on Computer Vision, pp. 868–884. Springer (2016)
15. Zheng, L., Shen, L., Tian, L., Wang, S., Wang, J., Tian, Q.: Scalable person re-identification: a benchmark. In: Proceedings of the IEEE International Conference on Computer Vision, pp. 1116–1124 (2015)
16. Zheng, Z., Zheng, L., Yang, Y.: Pedestrian alignment network for large-scale person re-identification. arXiv preprint arXiv:1707.00408 (2017)
17. Zhong, Z., Zheng, L., Cao, D., Li, S.: Re-ranking person re-identification with k-reciprocal encoding. In: 2017 IEEE Conference on Computer Vision and Pattern Recognition (CVPR), pp. 3652–3661. IEEE (2017)

Half Profile Face Image Clustering Based on Feature Points

Grzegorz Sarwas[(✉)] and Sławomir Skoneczny

Institute of Control and Industrial Electronics, Warsaw University of Technology,
Warsaw, Poland
{sarwasg,slaweks}@ee.pw.edu.pl

Abstract. In this paper the problem of hierarchical half profile face image clustering is considered. In order to solve this problem the computer vision methods based on a different local feature detectors like: Harris, BRISK, SURF, SIFT and FSIFT have been examined. For image clustering task the agglomerative hierarchical clustering procedure based on a dissimilarity matrix have been used. The achieved results have been compared to each other.

1 Introduction

The problem of face clustering is very important in image processing and computer vision fields. Recognition of people plays a crucial role in many security systems. In literature we can find dozens of papers devoted to methods of biometric identification [7]. Furthermore, Internet and in particular social media generate a lot of images presenting not only politicians or celebrities but also many photos of ordinary people. Recognition of a human face can help in the process of image searching and information classification. Nowadays, systems for analysing the content of a paper are based not only on the text description but also on the images. Knowledge about person presented on image can help in the process of classifying or clustering of given information.

In general, cluster analysis method (aka. clustering) is an algorithm that links similar objects into groups called clusters. In order to find the groups of similar objects it is necessary to define the similarity or dissimilarity measure on which depends the process of combining objects into the right groups. The hierarchical clustering is a recursive algorithm which starts by treating each observation as a separate cluster connecting the most similar objects in one new cluster. This process is continued until all the clusters are merged together into one group.

A problem of face clustering is not a trivial task. Profile face detection, recognition or grouping usually is different from the case when while processing the frontal views of human faces. Change of appearance, ageing process or even different facial expressions makes this problem quite complicated. Nowadays there is a lot of different techniques for face clustering. In literature we can find solutions based on Bayesian approach [11] or Hidden Markov Random Fields [16]. In [13] Schroff et al. proposed deep neural network solution called FaceNet

© Springer Nature Switzerland AG 2019
M. Choraś and R. S. Choraś (Eds.): IP&C 2018, AISC 892, pp. 140–147, 2019.
https://doi.org/10.1007/978-3-030-03658-4_17

for finding features for a face clustering. Another group of algorithms are based on local features [1,3].

In this paper the authors are focused on the process of half profile face clustering. In order to solve this problem methods based on a local feature detector are examined. A combination of the BRIEF, SURF and SIFT local descriptors with different feature point detectors like Harris, BRIEF, SURF, SIFT and FSIFT are compared. For image clustering the agglomerative hierarchical clustering [5] based on a dissimilarity matrix has been used.

The paper is organized as follows. In Sect. 2 feature points detectors are described. In Sect. 3 the proposed approach based on a hierarchical clustering and dissimilarity matrix created by using features detectors is presented. In Sect. 4 experimental results are presented and discussed. In the last section the article is summarized and concluded.

2   Feature Points Detectors

In this section the feature points detectors examined for half profile face clustering are described.

2.1   Harris Corner Detector

One of the important feature points are corners, which are interpreted like the points of linkage of two edges. In the computer vision applications the most often used corner detector and a fundamental method is Harris Corner Detector [4], which is defined like a mathematical operator invariant by rotation, scale and illumination. Its algorithm is based on the autocorrelation of image grayimage intensity values or image gradient values. The gradient covariance matrix is define as [9]:

$$G_{x,y} = \begin{bmatrix} \left(\frac{\partial I}{\partial x}\right)^2 & \frac{\partial I}{\partial x}\frac{\partial I}{\partial y} \\ \frac{\partial I}{\partial x}\frac{\partial I}{\partial y} & \left(\frac{\partial I}{\partial y}\right)^2 \end{bmatrix} = \begin{bmatrix} I_x^2 & I_x I_y \\ I_x I_y & I_y^2 \end{bmatrix}, \tag{1}$$

where I_x and I_y are the image gradients in the direction x and y, respectively. As the corner Harris Operator considers minimum α and maximum β eigenvalues of matrix $G_{x,y}$. A corner exists when two eigenvalues are large and similar in magnitude. It can be measured using determinant and trace of the matrix $G_{x,y}$. The measure of corner response R is defined as follows:

$$R = \alpha\beta - k(\alpha + \beta)^2 = det(G_{x,y}) - k(trace(G_{x,y}))^2, \tag{2}$$

where k is an empirically determined constant $k \in [0.04, 0.06]$. A corner point depends on R value, in the following way:

$$R > 0, \text{ for corner point,}$$
$$R \cong 0, \text{ for edge,}$$
$$R < 0, \text{ for flat region.}$$

2.2 BRISK

The BRISK is very effective method and often achieves comparable quality of matching to SURF algorithm at significantly reduced computation time [6]. BRISK uses scale-space approach to keypoint detection and for purpose of keypoint description it utilizes a special sampling pattern. This sampling pattern consisting of points lying on appropriately scaled concentric circles is applied (at the neighbourhood of each keypoint) to retrieve gray values.

The oriented BRISK sampling pattern is finally used to get pairwise brightness comparison results, which are assembled in the BRISK descriptor.

A classical binary descriptor is composed of three parts:

1. A sampling pattern: where the sample points are in the region around the descriptor.
2. Orientation compensation: some mechanism to measure the orientation of the keypoint and rotate it to compensate for rotation changes.
3. Sampling pairs: which pairs to compare when building the final descriptor.

The BRISK descriptor is composed as a binary string by concatenating the results of simple brightness comparison tests. Each keypoints has identified the characteristic direction what allow for orientation-normalized and hence achieve rotation invariance.

2.3 SIFT and SURF

The SIFT algorithm proposed by Lowe in 2004 consists of five main steps [8]:

1. Detection of scale-space extrema,
2. Keypoint localization,
3. Orientation assignment,
4. Keypoint descriptor,
5. Keypoint matching.

In order to detect keypoints with different scale the scale-space filtering based on the Laplacian of Gaussian (LoG) with the various variance of Gaussian distortion σ is applied. LoG with different values of the σ parameter detects blobs of various sizes. However, the LoG operator has quite large computational burden, so the SIFT method use Difference of Gaussians (DoG) as a LoG approximation. The DoG is obtained as the difference of Gaussian blurring of an image with two different σ's. These two (σ's must be significantly different (the first is k-times larger than the second one). This process is conducted for different octaves of the image in the Gaussian Pyramid.

Next, the local extrema over scale and space by comparing 8 neighbours pixels with 9 pixels in the next scale and the previous scale are searched. The found local extreme is a potential keypoint.

After that, localizations obtained by using the Taylor series expansion are refined to get more accurate results. If the extreme intensity is lower then a threshold value then this particular point is rejected. Moreover, low-contrast key-points and edge keypoints are eliminated so only strong interest points remain.

SURF detects feature points in a slightly different way than SIFT does. While the SIFT algorithm finds extrema in a $3 \times 3 \times 3$ neighbourhood, the SURF uses the Fast-Hessian detector to define the characteristic points [2].

2.4 FSIFT

FSIFT is the solution presented in [12]. Its main idea was based on the SIFT algorithm but the computing of DoG pyramid is replaced by calculating frac-tional derivative. In this paper we used the Riemann-Liouville (R-L) formula for calculating in time domain the fractional order differ-integral of $f(x)$ function expressed as:

$$_aD_x^\alpha f(x) = \frac{1}{\Gamma(n-\alpha)} \frac{d^n}{dx^n} \int_a^x \frac{f^{(n)}(\tau)}{(x-\tau)^{\alpha-n+1}} d\tau \tag{3}$$

where $\alpha \in \mathbb{R}$ is a fractional order of the differ-integral of the function $f(x)$ and for $n \in \mathbb{N} \cup \{0\}$ we have:

$$n - 1 < \alpha \leq n \quad \text{for} \quad \alpha > 0,$$
$$n = 0 \quad \text{for} \quad \alpha \leq 0.$$

The Fourier transform of the Riemann-Liouville fractional derivatives with the lower bound $a = -\infty$ is [10]:

$$_{-\infty}D_x^\alpha f(x) = \frac{1}{\Gamma(n-\alpha)} \int_{-\infty}^x \frac{f^{(n)}(\tau)}{(x-\tau)^{\alpha+1-n}} d\tau =_{-\infty} D_x^{\alpha-n} f^{(n)}(x) \tag{4}$$

and we assume here that $n - 1 < \alpha < n$ and $n \in \mathbb{N}$. After applying the Fourier transform to Eq. (4) we arrived at:

$$\mathcal{F}(D^\alpha f(x)) = (j\omega)^\alpha F(\omega). \tag{5}$$

Therefore the pair of fractional derivative in the time domain and in the fre-quency domain is:

$$D^\alpha f(x) \leftrightarrow (j\omega)^\alpha F(\omega), \quad \alpha \in \mathbb{R}^+. \tag{6}$$

For any two-dimensional function $g(x, y)$ absolutely integrable in $(-\infty, \infty) \times (-\infty, \infty)$ the corresponding 2-D Fourier transform is as follows [10]:

$$G(\omega_1, \omega_2) = \int g(x, y) e^{-j(\omega_1 x + \omega_2 y)} dx dy. \tag{7}$$

Therefore we can write the formula for fractional-order derivatives as:

$$D_x^\alpha g = \mathcal{F}^{-1}\left((j\omega_1)^\alpha G(\omega_1, \omega_2)\right),$$
$$D_y^\alpha g = \mathcal{F}^{-1}\left((j\omega_2)^\alpha G(\omega_1, \omega_2)\right), \tag{8}$$

where denotes $\mathcal{F}^{-1}$ is the inverse 2-D continuous Fourier transform operator. This formula have been extensively used in FSIFT algorithm.

The Fractional Scale-Invariant Feature Transform (FSIFT) algorithm presented in 2017 is an approach to keypoint detection where the step of computing DoG pyramid has been replaced by calculating fractional derivative. This method can be demonstrated in the following steps:

1. Build the classical Gaussian Pyramid.
2. For each scale calculate the collection of images created by fractional order derivative with the following orders of $\alpha = 1.75, 1.85, 1.95, 2.05, 2.15, 2.25$.
3. For derivative images search for extrema over order and space.
4. The interesting points' locations are updated by using an interpolate based on the second-order Taylor-series.
5. Reject all extrema with low contrast.
6. Refine keypoint locations neglecting points that lie on edges.

3   Hierarchical Clustering

Hierarchical clustering is one of the most popular methods of object grouping that organizes data. An agglomerative clustering method starts by treating each observation as a separate cluster. Then, in every step two closest clusters are joined in a new one. This iterative algorithm works until all data is combined into one cluster.

In the presented research a concept of hierarchical face clustering based on the solution suggested by Antonopoulos et al. [1] using a dissimilarity matrix is performed. This matrix is transformed into the previously defined number of clusters by using an agglomerative hierarchical clustering method. To create dissimilarity matrix for half profile face clustering we propose FSIFT feature point detector. The clustering effectiveness of this algorithm has been compared with the Harris-SURF, BRISK, SURF and SIFT feature detectors and descriptors used for building a dissimilarity matrix.

3.1   Building the Dissimilarity Matrix Using Keypoint Detectors

The dissimilarity matrix $\mathbf{D}$ is square and symmetric. It contains all pairs of differences between samples that should be grouped. The size of this matrix is $N \times N$, where N is the total number of clustering face images. Each element D_{ij} defines the dissimilarity between facial images A_i and A_j. This dissimilarity is defined by the following formula [1]:

$$D_{ij} = D_{ji} = 100 \left(1 - \frac{M_{ij}}{min(K_i, K_j)} \right), \tag{9}$$

where M_{ij} is the maximum number of keypoints matches between the pairs (A_i, A_j), (A_j, A_i) and K_i, K_j are numbers of keypoints found in A_i, A_j, respectively.

$D_{i,j}$ is in a range of $[0, 100]$ and higher values of $D_{i,j}$ imply higher dissimilarity between images A_i and A_j. The feature points are matched by using the Euclidean distance between feature vectors. The match is valid if the distance between the considered pixels and the closest neighbour is less than the distance between the considered pixel and the second closest match.

4 Experiments

Two types of experiments have been performed on the set of 12 human faces in the left half profile from the IMM Face Database [15]. The first one conducted on the original images and the second on images sharpened by the LUM sharpener [14]. Some local keypoints detectors like: Harris, BRISK, SURF, SIFT and FSIFT have been used. Two image resolutions have been applied $(512 \times 512, 150 \times 150)$. The examples of the detected keypoints detected by using implemented algorithms are shown in Fig. 1. The results of conducted clustering experiments are presented in the Table 1.

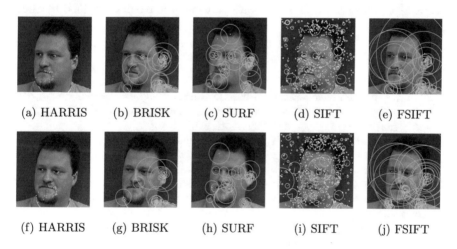

(a) HARRIS   (b) BRISK   (c) SURF   (d) SIFT   (e) FSIFT

(f) HARRIS   (g) BRISK   (h) SURF   (i) SIFT   (j) FSIFT

Fig. 1. All used descriptors of 2 half profile example images.

Table 1. Clustering error for 12 classes and 48 images.

Sharpening	Resolution	HARRIS	BRISK	SURF	SIFT	FSIFT
–	512×512	0.6667	0.5833	0.1250	0.0833	0.0833
LUM	512×512	0.7917	0.2708	0.0417	0.0417	0.0833
–	150×150	0.6250	0.3750	0.2500	0.2500	0.0833
LUM	150×150	0.4792	0.6458	0.2917	0.2500	0.0000

In the first case, SIFT and SURF algorithms were the best ones, while FSIFT were comparable but it gave a little bit worst results. However, for 150×150 pixels images FSIFT won the contest (0% error) although SIFT and SURF methods were very successful. BRISK method was significantly worst and Harris detector gave completely unacceptable results. Sometimes LUM sharpener helps to increase the proper rate of keypoints detection and therefore we obtained almost perfect clustering results.

5 Conclusions

In this paper the problem of half profile face images clustering is addressed. For image clustering the agglomerative hierarchical clustering based on a dissimilarity matrix was used. The dissimilarity matrix was build by using local feature matching based on a FSIFT feature points. Achieved results have been compared with another local feature point detectors like Harris, BRISK, SURF and SIFT. For small resolution images FSIFT obtained the best results, while for higher resolution face images it was comparable with SURF and SIFT algorithms. The BRISK algorithm appeared to by quite poor but the results obtained by Harris corner detector were completely unacceptable.

References

1. Antonopoulos, P., Nikolaidis, N., Pitas, I.: Hierarchical face clustering using SIFT image features. In: 2007 IEEE Symposium on Computational Intelligence in Image and Signal Processing, pp. 325–329, April 2007
2. Bay, H., Ess, A., Tuytelaars, T., Gool, L.V.: Speeded-up robust features (SURF). Comput. Vis. Image Underst. **110**(3), 346–359 (2008). Similarity Matching in Computer Vision and Multimedia
3. Geng, C., Jiang, X.: SIFT features for face recognition. In: 2009 2nd IEEE International Conference on Computer Science and Information Technology, pp. 598–602, August 2009
4. Harris, C., Stephens, M.: A combined corner and edge detector. In: Proceedings of Fourth Alvey Vision Conference, pp. 147–151 (1988)
5. Larsen, B., Aone, C.: Fast and effective text mining using linear-time document clustering. In: Proceedings of the Fifth ACM SIGKDD International Conference on Knowledge Discovery and Data Mining, KDD 1999, pp. 16–22. ACM, New York (1999). https://doi.org/10.1145/312129.312186
6. Leutenegger, S., Chli, M., Siegwart, R.Y.: BRISK: binary robust invariant scalable keypoints. In: 2011 International Conference on Computer Vision, pp. 2548–2555, November 2011
7. Liao, S., Jain, A.K., Li, S.Z.: Partial face recognition: alignment-free approach. IEEE Trans. Pattern Anal. Mach. Intell. **35**(5), 1193–1205 (2013)
8. Lowe, D.G.: Distinctive image features from scale-invariant keypoints. Int. J. Comput. Vis. **60**(2), 91–110 (2004)
9. Malik, J., Dahiya, R., Sainarayanan, G.: Harris operator corner detection using sliding window method. Int. J. Comput. Appl. **22**(1), 28–37 (2011)
10. Podlubny, I.: Fractional Differential Equations. Academic Press, San-Diego (1999)

11. Prince, S.J.D., Elder, J.H.: Bayesian identity clustering. In: 2010 Canadian Conference on Computer and Robot Vision, pp. 32–39, May 2010
12. Sarwas, G., Skoneczny, S., Kurzejamski, G.: Fractional order method of image keypoints detection. In: 2017 Signal Processing: Algorithms, Architectures, Arrangements, and Applications (SPA), pp. 349–353, September 2017
13. Schroff, F., Kalenichenko, D., Philbin, J.: FaceNet: a unified embedding for face recognition and clustering. In: 2015 IEEE Conference on Computer Vision and Pattern Recognition (CVPR), pp. 815–823, June 2015
14. Skoneczny, S.: Contrast enhancement of color images by nonlinear techniques. Prz. Elektrotech. R $86(1)$, 169–171 (2010)
15. Stegmann, M.B., Ersbøll, B.K., Larsen, R.: FAME - a flexible appearance modelling environment. IEEE Trans. Med. Imaging $22(10)$, 1319–1331 (2003)
16. Wu, B., Zhang, Y., Hu, B.G., Ji, Q.: Constrained clustering and its application to face clustering in videos. In: 2013 IEEE Conference on Computer Vision and Pattern Recognition, pp. 3507–3514, June 2013

Communications and Miscellaneous Applications

Software-Defined Automatization of Virtual Local Area Network Load Balancing in a Virtual Environment

Artur Sierszeń[1]([✉]) and Sławomir Przyłucki[2]

[1] Lodz University of Technology, Zeromskiego 116, 90-924 Lodz, Poland
artur.sierszen@p.lodz.pl
[2] Lublin University of Technology, Nadbystrzycka 38A, 20-618 Lublin, Poland
http://www.p.lodz.pl
http://www.pollub.pl

Abstract. Nowadays, data centre operators have to implement new solutions quickly, providing their clients with high-quality services related to access to data or application. For this reason, open source solutions that enable full control over the development of a network environment are very popular. In this group, the leader is the solution of Cumulus Networks - Cumulus Linux. During their work on implementing automatic mechanisms of load balancing of devices working in the L2 ISO/OSI, the authors noticed the lack of such mechanisms in the Cumulus Linux solution. The article presents a solution to this problem. Tests have shown the usefulness of this solution.

Keywords: Software Defined Network · SDN · Automatization
VLAN · Virtual Local Area Network · Load balancing
Virtual environment · Cumulus Networks · STP
Spanning Tree Protocol · RSTP · Rapid Spanning Tree Protocol

1 Introduction

The variety of technical solutions of data centres in cloud environments presents a great difficulty in a selection of a suitable environment during implementation, maintenance, and development. The Gartner report from 2017 [1] contains information on the trend of moving away from proprietary solutions in data centres to open ones. (...) 75% of end users expect an increase in relevance of open networking in the next 24 months. This requirement is mostly unfulfillable by vendors, since the majority of the solutions considered in this research are proprietary. In the same report, Cumulus Network was presented as visionary in this field. Not only has it created an open network system dedicated to network switches (Open Network Operating System for bare metal switches) as well as virtual environments, but it has successfully implemented this solution for large networks. This is evidenced by the fact that 32% of companies from the Fortune 50 list use Cumulus Linux in their data centres.

© Springer Nature Switzerland AG 2019
M. Choraś and R. S. Choraś (Eds.): IP&C 2018, AISC 892, pp. 151–160, 2019.
https://doi.org/10.1007/978-3-030-03658-4_18

The above-mentioned facts encouraged the authors to choose the Cumulus Linux environment for research related to the automation of network solutions based on the Software Define Network. They focus on:

– the aspect of equal load of devices in data centres and/or server rooms [2],
– automation of configurations in order to develop self-steering solutions [3].

Both issues are very important in the aspect of maintaining the infrastructure of data centres. The issue of network load-balancing is still relevant to modern computer networks. The use of many redundant devices in computer networks contributes to network reliability. A network constructed in this way may function even when a part of links or devices fail. Load-balancing mechanisms in such topologies even allow the use of all available network resources. The aspect of implementing mechanisms that balance the use of devices should be taken into account as early as at the time of network design, which, in future, enables the already implemented computer network to prevent unnecessary data transmission by devices. Lack of mechanics balancing network load can lead to a disproportion in the load of main switches for individual VLAN networks and may contribute to problems related to power supply and cooling.

1.1   Spanning Tree Protocol in Cumuls Linux

Configuring redundant connections in a switched network increases the reliability of connections between devices. However, this creates additional problems (e.g.: broadcast storm, MAC address table overflow/mismatch etc.). To solve them, the Spanning Tree Protocol (STP) was developed [4].

This protocol creates a graph (STP tree) without a loop and establishes backup links. During normal operation, the network blocks them so that they do not transmit any data. Only one path is used for data exchange, upon which communication can take place. On the top of the graph is the main switch called the Root Bridge (RB) that manages the network. The choice of the RB switch is based on an identifier configured by the network administrator. When the STP detects a problem, such as a broken link, it reconfigures the network by activating a backup link.

The STP has the disadvantage of a long time of reaching network convergence, often in excess of 60 s. Therefore, many modifications have been made, including:

– Rapid Spanning Tree Protocol (RSTP) which provides shorter recovery time for disaster recovery (standard IEEE 802.1w),
– Multiple Spanning Tree Protocol (MSTP) which enables load balancing and improves fault tolerance by providing multiple paths for data traffic (IEEE 802.1s and its IEEE 802.1Q-2005 modification),
– Shortest Path Bridging (SPB) released as the IEEE 802.1aq standard in May 2012, which enables multi-path routing.

The largest vendors of networking devices have also developed many of their own alternatives, e.g. Cisco Corporation (Per-VLAN Spanning Tree (PVST), Per-VLAN Spanning Tree Plus (PVST+), Multiple Instance Spanning Tree Protocol (MISTP)), or Juniper Networks (VLAN Spanning Tree Protocol (VSTP)).

In the Cumulus Linux environment, PVST and PVRST are implemented. Both form a spanning tree for each VLAN network separately (PV means per VLAN), for the STP and RSTP protocol respectively.

This is a big drawback, according to the authors. If e.g. 200 VLANs are configured in a single network, the root bridge will have to support up to 200 separate spanning trees. A solution may consist in dividing this task between a few other switches that will act as root bridges for different VLANs, which is not difficult in terms of network administration. However, this is a static rewrite and it does not take the actual traffic volume into account, i.e. the amount of data served by individual VLAN networks, therefore the tree spanning this network.

2   Background and Related Works

The authors are considering two solutions to that issue. The solutions are based on:

- an external server collecting information on the volume of incoming data in separate VLANs and sending appropriate reconfiguring commands to switches based on the analysis of the collected data,
- additional software implemented on a switch with no need to install an external server.

The first solution has been previously tested [2] and its laboratory implementation is shown in Fig. 1. It has a monitoring station connected to the x1 switch with software that collects information about the amount of traffic carried by individual VLANs. This data was obtained using SNMP, NetFlow/J-Flow, TCL and bash scripts etc. The use of script languages on monitoring devices makes the proposed solution independent from other protocols but requires that certain command interpreters be added in the reconfiguring mechanism.

On the server, a script was enabled which compared the quantity of data for Root Bridges within STP bridges. If the total quantity of data for all packages belonging to the VLAN of a single region was higher by 20%, the VLANs were re-assigned to regions. The application mechanism put VLANs in order in terms of the volume of transferred data and assigned them, one by one, alternatively, to one out of two regions. To start the reconfiguration, it was enough to send an appropriate command to both Root Bridges.

Although it was a solution to the problem, this mechanism had several disadvantages:

- the central monitoring and management server was a potential single point of failure,
- there was a need to modify the control application for specific device configuration files,

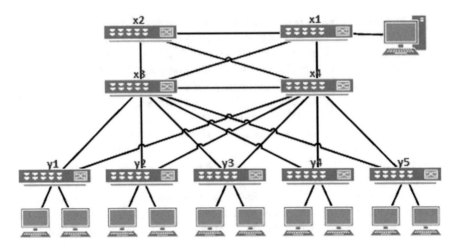

Fig. 1. Test Lab topology.

– the amount of administrative work is increased, which translates into the additional cost of maintaining and updating the server and the development environment,
– you have to pay the purchase costs of the server and its software.

These defects forced the authors to develop a mechanism that will be installed directly on switches.

3   Proposed Solution

The proposed solution is designed for typical data centre networks, with switches at distribution/core and access layer work.

In order to distribute the amount of data traffic carried by VLANs and handled by a single Root Bridge evenly, it was decided to develop a software solution implemented directly on switches.

The following parameters and functionalities have been defined:

1. Easy to run on individual devices;
2. Openness and transparency of the solution;
3. Security of the solution;
4. Minimum management;
5. Self-configuration mechanism (automation).

Regarding Point 1. All parts of the solution are included in a single file.
Regarding Point 2. Cumulus Linux supports a few programing languages (*Bash, Ruby, Perl, Python*) by default. The authors have decided to use the Python environment (it is very popular among network administrators), and well-known

libraries (like *paramiko 2.0.2* (SSH library) or *pyp2p 0.8.2* (Python P2P networking library)). Both of these factors should facilitate checking the code before implementation in a test environment and make the solution safer.

Regarding Point 3. All data about the volume of traffic and all reconfiguration commands are sent over a SSH connection.

Regarding Point 4. The solution is run only once by the network administrator (on switches chosen as Root Bridges), without any configuration parameters. For correct operation, it is necessary to configure the management network and to enable SSH traffic and SSH logging on devices.

Regarding Point 5. All parameters necessary to run and maintain of solution are received automatically. Of course, the administrator can define his own settings by overriding the default parameters.

3.1   Description of the Solution

The presented solution consists of three segments:

1. Management of connections between switches (peers), which are Root Bridges (solutions can only be installed on them),
2. Exchange of data regarding the amount of traffic carried by individual VLANs,
3. Mechanism of load balancing of an individual Root Bridge.

Managing Connections Between Switches (Peers). It was decided to use peer-to-peer technology to develop a model of connections between switches that act as the root bridge.

- In the first stage (after launching the program), the broadcast about the appearance of a new node (Discovery Packet - Broadcast) is sent in the management network. This message also informs about the IP address from the management network, supported by the node, and the MAC address of the management interface (eth0).
- The other nodes (as unicast) send analogous information (their IP address and MAC address) (Discovery Packet - Unicast).
- All nodes are connecting using the SSH connection. There is a default account and password for the SSH support, but for security reasons it is recommended to change it. A MESH all-to-all network is created.
- Every 20 s a message is sent (already through the SSH connection) that the solution (application) is still running on the Keepalive Packet.
- If the neighbour does not receive the Keepalive package within 45 s, information is sent to all other nodes about the lack of communication with the node concerned (Delete-Peer Packet). The node is removed from the load balancing functionality if all nodes confirm the lack of communication.

In the manner described above, a network of connections between all nodes that perform the Root Bridge role is provided.

Exchange of Data. Every 120 s, information about the total data traffic within each VLAN is collected by individual switches. This data is sent to all other nodes. It is stored for 1800 s, and after that time it is removed from the logs (there is an option to archive it by sending to an external server through SCP (Secure copy). Storing this data allows to illustrate the entire data distribution within all Route Bridge switches. The authors chose both time parameters based on laboratory tests. Nevertheless, in future work it is recommended that these parameters be analysed and optimized.

Mechanism of Load Balancing. Based on the MAC address of the management interface, the order of the switches in the load balancing process is determined; they are sorted from the smallest to the largest address. Theoretically, MAC addresses are globally unique, but there is a possibility of their substitution. Therefore, if the system detects two identical MAC addresses, both switches are deleted from the list and do not participate in the process of load balancing (Delete-Peer Packet is sent to both switches).

The load balancing process itself is based on the analysis of data collected from all switches

- In the first stage, the loads within each Root Bridge switch are sorted, starting from those responsible for generating the largest volume of traffic. The analysis covers received by the switch in 5 subsequent updates (i.e. for the last 600 s).
- Information is collected about the entire volume of data traffic within the entire network (sum of all VLANs).

Depending on the number of switches in the system, a different detection threshold has been assumed for reconfiguring the rewriting of the VLAN to individual Root Bridge. These data is collected in Table 1.

Table 1. The dependence between the number of switches and the triggering of the load balancing process

Number of root bridges	Threshold [%]
2	10
3	6
4	5
5	4
6	3
More than 6	2

If the difference between the most and the least loaded switch exceeds the indicated threshold, the configuration mechanism is started. It consisted of scheduling the VLANs in terms of the volume of data transferred and assigning them one by one to each Root Bridge switch.

The authors consider further use of more advanced algorithms.

4 Test Scenario, Evaluation and Results

4.1 Testing Environment

In order to verify the solution, a research environment based on iESX 6.5 was built. The tested topology included 9 Cumulus Linux VX 3.5.0 switches – Fig. 2:

1. 4 switches in the distribution layer that could act as root bridges (CumulusVX 1 to 4),
2. 5 switches in the access layer acting as the main switches of the access layer (CumulusVX A to E).

For each switch operating in the access layer, two layer 2 switches were added along with 20 stations containing the traffic generator - *Ostinato Network Traffic Generator*. *Ostinato* has a client and a server mode where servers can be run on multiple hosts and the client can connect to multiple servers. This allows a single client to manage ports present in multiple servers. This functionality allowed to configure up to 15 connections/traffic sources (with different vlan tagging) on a single host machine. In total, traffic was generated from 288 sources – this number has been dictated by limitations resulting from the amount of the operating memory of the test environment (it was the maximum number of sources whose work did not increase the total resources by more than 1%.)

4.2 Test Scenario

In order to verify whether the solution operates correctly, the network is configured as follows:

- 36 VLANs were defined - so that there are 8 sources generating data traffic for each VLAN.
- The vlan networks have been configured in such a way that a single distribution switch (Cumulus A to E) does not support more than 10 different VLANs.
- 10 different traffic patterns have been defined in the Ostinato Network Traffic Generator, which were randomly selected and run with a random delay (from 30 to 1000 s) – this guaranteed a variety of traffic volumes within each VLAN.

The solution presented above (Sect. 3) was run only on distribution switches (Cumulus 1 to 4). The following start-up sequences have been introduced:

1. on each of the 4 switches, one after the other (with different delay, which simulated adding a new switch)

158   A. Sierszeń and S. Przyłucki

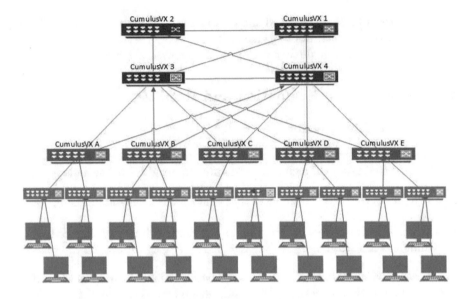

Fig. 2. ESXi Tested Lab topology.

2. simultaneously on four switches (running at the same time – time control via Time Network Protocol)
3. on two randomly selected switches (running at the same time)
4. on two randomly selected switches (with different delay)
5. on three randomly selected switches (running at the same time)
6. on three randomly selected switches (with different delay)

Each sequence was repeated at least 20 times.

4.3   Results

The conducted tests, which consisted in changing the quantity of data generated in separate VLANs, confirmed that the application works correctly. The mechanism of load balancing regions assigned separate VLANs correctly to the quantity of data transferred within the Root Bridge. Its operation could be checked in detail owing to the results from the monitoring station.

The quantitative distribution of the VLANs between all switches in a given test sequence worked in a way that was expected (Table 2).

The volume of traffic carried by a VLAN belonging to a single root bridge shows that the load has been distributed evenly among the switches. This is particularly evident when compared to the original STP concept. When we decide to use the original PVST or PVRST without the proposed extensions (in the same test environment), all 36 VLANS belong to a single Root Bridge. It means that this switch supports virtually 100% of the transferred traffic and the other 3 switches have a total of less than 0.5%.

Table 2. Result of test (*data from all test).

Number of switches in test sequences	Number of VLANs assigned to a single Root Bridge	Minimum value of traffic represented by single Root Bridge [% of all traffic]*	Maximum value of traffic represented by a single Root Bridge [% from all traffic]*
4	9 VLANs	22%	27%
3	12 VLANs	28%	37%
3	18 VLANs	45%	57%

5 Conclusion and Future Work

The tests also showed the need for further work related to the improvement of the network switching algorithm between individual switches acting as Root Bridge.

Further work is planned to implement and test other load balancing algorithms, e.g. Opportunistic Load Balancing (OLB) and Load balance Min-Min (LBMM). The authors are also aware of the need to conduct tests in the data centre network with a real (not generated) exchange of frames in the second ISO/OSI layer.

Despite this, the authors believe that the solution to the lack of a load-balancing mechanism in Cumulus Linux Per Vlan Spanning Tree Protocol may translate into more effective use of hardware, in particular in campus networks. That may lead to the reduction of operation costs (reduced load of cooling systems in server rooms, balanced consumption of electricity) as well as costs of network modernization (no need to replace devices; a software update is only needed in case of a solution based on the implementation of additional software to a switch).

Acknowledgements. This work was financed by the Lodz University of Technology, Faculty of Electrical, Electronic, Computer and Control Engineering as a part of the statutory activity (project no. 501/12-24-1-5438).

References

1. Ciscato, D., Mark Fabbi, M., Andrew Lerner, A.: Gartner Report - Gartner Quadrant for Data Center Networking, 3 July 2017
2. Sierszen, A., Przyluski, S.: Load balancing regions mechanism in multiple spanning tree protocol. Stud. Inform. **38**(3), 114–124 (2017)

3. Wojciechowski, R., Sierszen, A., Sturgulewski, L.: Self-configuration networks. In: Conference: 8th International Conference on Image Processing and Communications (IP and C) Location: UTP Univ Sci & Technol. Image Processing and Communications Challenges 7 Book Series: Advances in Intelligent Systems and Computing, Bydgoszcz, Poland, 09–11 September 2015, vol. 389, pp. 301–308 (2016)
4. Perlman, R.: An algorithm for distributed computation of a spanning tree in an extended LAN. In: ACM SIGCOMM Computer Communication Review 1985, vol. 15, no. 4, pp. 44–53. ACM, New York (1985)

Modified TEEN Protocol in Wireless Sensor Network Using KNN Algorithm

Abdulla Juwaied$^{(\boxtimes)}$, Lidia Jackowska-Strumiłło, and Artur Sierszeń

Institute of Applied Computer Science, Lodz University of Technology,
ul. Stefanowskiego 18/22, 90-924 Łódź, Poland
{ajuwaied,asiersz}@iis.p.lodz.pl, lidia_js@kis.p.lodz.pl

Abstract. Nowadays Wireless Sensor Network has become an interesting filed for researchers. The main concerns about Wireless Sensor Network are the life time and energy consumption. Because of these two reasons new protocols improving these network features are being developed all the time. This paper studies TEEN (*Threshold Sensitive Energy Efficient Sensor Network*) protocol. We improve the performance of choosing the Cluster Heads in the cluster areas using KNN (*K Nearest Neighbours*) algorithm, which use a distance between the nodes as a similarity measure to classify the nodes in the network. Therefore, most of routing protocol techniques and sensing task require to check the position of sensors in network to obtain information of location with high accuracy sensors in network. The modification in this protocol improves the life time and energy consumption compared with the original TEEN protocol.

Keywords: Cluster Head · Modified TEEN protocol
Wireless Sensor Network · Energy

1 Introduction

The main goal of using Wireless Sensor Network is to create a connection to the real world through sending and receiving information from the sensors (Fig. 1). The infrastructure of WSN consist of two main parts, the *data dissemination* network and the *data acquisition*. The *data dissemination* network is a combination of wireless network and wired network that provides information of the data acquisition in user network and the second part is acquisition network which contains cluster area, nodes and base station [2]. All the sensors in Wireless Sensor Network are autonomous, not controlled by the user. The sensors in the WSN are constrained in terms of energy consumption and battery life. All the sensors in WSN consist of sensing, power, processing and transaction units.

There are two main architectures of Wireless Sensor Networks: *hierarchical* and *flat*. In the *hierarchical* architecture the network is divided into groups of nodes or cluster areas, which specify goals of the sensors in the network, on the other hand in the *flat* architectures all nodes in the network contribute in the decision-making process and participate in the internal protocol [3].

© Springer Nature Switzerland AG 2019
M. Choraś and R. S. Choraś (Eds.): IP&C 2018, AISC 892, pp. 161–168, 2019.
https://doi.org/10.1007/978-3-030-03658-4_19

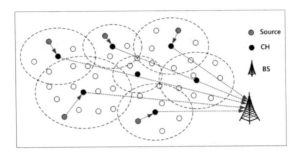

Fig. 1. Typical Wireless Sensor Network; where: CH - Cluster Heads, BS - Base Station.

2   Sensors in WSN

Most of routing techniques and sensing tasks require to check sensor position [11,12] in the network to obtain information about location of the sensors in the network with high accuracy. Sensor networks can be classified into proactive and reactive networks depending on the mode of functioning and type of target application [4].

Proactive network: in this scheme the sensors periodically switch on their nodes and transmitter to sense the environment first then transmit the data. This type of network is suited for application requiring periodic data monitoring.

Reactive network: the sensors in this network react immediately to sudden and drastic changes in the value of sensed attribute. This type of network is suited for time critical application.

3   TEEN (Threshold Sensitive Energy Efficient Sensor Network) Protocol

TEEN protocol is based on cluster hierarchical approach and use data centric method. This type of protocol can be classified as a reactive protocol because sensors in the network are arranged in hierarchical clustering [4]. Therefore, The Cluster Heads (CH) in the network send acknowledge to higher level of Cluster Heads (first and second level) until the information reach the Base Station (BS).

Nodes clustering in TEEN protocol is shown in Fig. 2. As it was mentioned previously, this protocol uses hierarchical clustering and each cluster area has 1 cluster head to collect the data from the sensors and then to send the information to the Base Station or to the 2nd level cluster head. The white dotes represent nodes, they connect with CH in the 1st level, then these CHs connect with other CHs in 2nd level.

3.1   TEEN Functionality

When the Cluster Head is selected, the user sets its attributes. When the CH receives these attributes (Thresholds) they will be broadcasted to all nodes in cluster area.

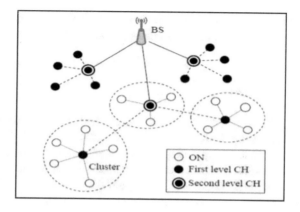

Fig. 2. Hierarchical clustering topology in TEEN protocol [1].

The functionality of TEEN is:

Hard threshold (H_T): This is a threshold value for the sensed attribute. It is the absolute value of the attribute above which, the node sensing this value must switch on its transmitter and report it to its CH [4,5].

Soft threshold (S_T): This is a minimum change of the sensed attribute value which triggers on the sensor's transmitter [4,5].

The nodes sense their environment continuously in the network, then transmit the sensed attribute value only when it exceeds the *Hard threshold* (H_T). *Sensed value* (S_V) is an internal variable which is used to store the transmitted sensed value. The node senses data again and when the sensed value exceeds the *Soft threshold* (S_T), it starts transmitting data [9].

3.2   Main Feature of TEEN Architecture

- The cluster heads in the network need to perform additional processing on the data. Therefore, more energy should be preserved for them.
- Cluster heads at increasing levels in the hierarchy need to transmit data for very large distances.
- All nodes in the cluster area transmit only to their cluster head, because the energy consumption is lower then.

3.3   TEEN Properties

The data transmission can be easily controlled because as it was mentioned earlier there are two types of thresholds. That means that the energy used for transmission can be reduced. Therefore, TEEN is recommended for time critical applications for complementing reacting to large changes [5].

In the other hand, it is not suitable for periodic reports applications, because when the values of the attributes are below threshold, the user may not get any data at all. Also in TEEN some cluster heads are not in the range of communication, so the data may be lost, because the data in the network is controlled only by cluster heads [4,5,9].

4   Implementation

Unlike LEACH [10] and SEP Protocols [8], in the TEEN protocol the cluster heads broadcast two type of threshold: *Hard threshold* and *Soft threshold* to the node member to control the quantity of data transmission. The *Hard threshold* (H_T) and *Soft threshold* (S_T) base on the instance value which is manually set by the user. As previously mentioned, the main disadvantage of TEEN protocol is that when the threshold not reached, then all nodes in the cluster area will not have any connection with other nodes in the network. Therefore, the base station in the network will not have any information from these nodes and also if any sensor is died. From this point the base station will not know about the life time and energy consumption of CHs because of the died CHs in the network.

The implementation of the TEEN protocol with the choice of nodes, which should join alive CHs using KNN classification algorithm was done in the following steps:

– *First step*: Implement the original TEEN protocol using MATLAB simulation which is shown in Fig. 3, the field having coordinate size 100(X)*100(Y) meters, normal nodes (o), cluster heads (*), number of nodes in the field n = 50 and thresholds for transmitting data to the cluster heads $(H_T) = 50$ and $(S_T) = 5$. We will get the following implementation, the number of Nodes = 50 and number of CHs = 14. Only 4 nodes will be alive as a CHs and the rest will die. Therefore, there user will not get any information of nodes which are communicated with died CHs.
As was shown in Fig. 3 left, 10 cluster areas with red X symbol will not have any connection with user because of the died CHs. In this case the user will not able to know how many nodes are alive or dead in the network then the life time of the network will be unknown. Table 1 shows the number of alive and died CHs for the simulation of the original TEEN protocol.
– *Second step*: election a cluster heads for the nodes, Eq. (1) defines the probability threshold, which is used for each node in each round to check if this node becomes a CH. when the random number is smaller than T(n), the node decides to become CH.

$$T(s_i) = \begin{cases} \dfrac{P_i}{1-P_i\left(r \ mod \ \frac{1}{P_i}\right)} & \text{if } s_i \in G \\ 0 & \text{Otherwise} \end{cases} \tag{1}$$

where: $T(s_i)$ - threshold, P_i - probability of change of node to become a CH, r - current round number, G - is the set of nodes that will be chosen as CHs at each round r.
– *Third step*: checking if there is a dead node in the network before using KNN to choose nodes to join the cluster heads.

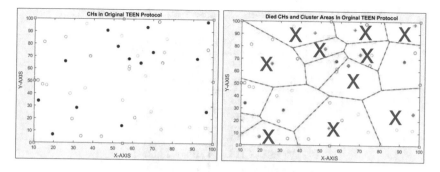

Fig. 3. Simulation of Original TEEN Protocol (left: CHs in Original TEEN Protocol, right: Died CHs in the network)

Listing 19.1. Checking dead nodes.

```
if (A(x).E<=0)
   plot(A(x).xa, A(x).ya,'blue .');
   dead=dead+1;
   if (A(x).Energy==1)
      Dead_a=Dead_a+1;
   end
   if (A(x).Energy==0)
            Dead_n=Dead_n+1;
         end
end
if A(x).E>0
   A(x).type='N';
   if (A(x).Energy==0)
      plot(A(x).xa,A(x).ya,'*');
   end
   if (A(x).Energy==1)
      plot(A(x).xa,A(x).ya,'o');
```

- *Fourth step*: After the selection of cluster heads is done, they try to send message to other nodes in the network to decide which CH to join. The nodes in the network are chosen for the nearest CH and for lower energy consumption of CHs using KNN algorithm. Table 2 show the number of live CHs and energy consumption for each CH in the network.
- *Fifth step*: All nodes in network start to join cluster heads using KNN algorithm using the position of nodes depending on the coordinate of X and Y, after the dead node is known for the user. The network will be divided into sub groups, which consist of set of nodes and one cluster head.

Table 1. Cluster heads dist. to base station in original TEEN

X	Y	No. CH	DIS-BS	CHs State
12.195	33.866	CH1	138.746	Live
25.595	65.699	CH2	125.392	Died
31.511	28.320	CH3	120.456	Live
46.920	90.720	CH4	110.831	Died
54.520	14.247	CH5	101.954	Died
52.252	77.898	CH6	101.651	Died
57.910	67.966	CH7	93.826	Live
57.910	67.966	CH8	93.826	Died
67.136	93.260	CH9	93.476	Died
63.933	64.508	CH10	87.281	Died
70.038	73.240	CH11	83.271	Died
97.215	97.484	CH12	71.000	Died
92.059	68.030	CH13	60.682	Live
96.862	25.521	CH14	58.505	Died

4.1 Implementation Results

In the modified TEEN protocol using KNN algorithm [6,7], the nodes are selected to chosen CHs, which are alive. In the presented example only 4 CHs are alive. Comparing to the original protocol more than 9 CHs are dead and the user will lose an information because of this. Therefore, in this paper we focused on checking the distance of nodes and energy first then all nodes can join CHs. It is shown in the Fig. 4, how the node is selected to join CHs using KNN algorithm. Table 2 show the number of CHs after modification. Comparing with other results dead CHs is one of the most drawback in this protocol.

Therefore, the network in modification were divided into four cluster areas without died cluster heads and all the node in the network will join one CH. Then the result had less amount of energy dissipation rate comparing with original TEEN protocol because the original protocol does not provide clearly information for user about dead nodes as it was shown in Fig. 3.

Table 2. Live cluster heads dist. to base station and energy after TEEN modification.

X	Y	No. CH	DIS-BS	Energy
57.91024	67.96573	#CH1	93.825	0.06571
31.51108	28.31958	#CH2	120.451	0.052174
12.19485	33.86611	#CH3	138.74	0.046396
92.05854	68.0299	#CH4	60.68	0.080969

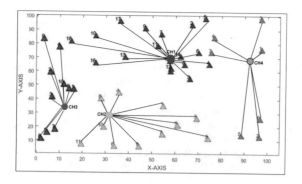

Fig. 4. Modification of TEEN using MATLAB - cluster heads (O) with belonging to them nodes.

5   Conclusion

This paper presents a modified TEEN routing protocol using KNN algorithm. In the original TEEN protocol there are some drawbacks, one of this disadvantage is when the *Hard threshold* (H_T) is not checked correctly then the nodes will not be able to be connected together. Comparing with other results dead CHs are the most drawback in this protocol. Therefore, the user will not know about the dead cluster head nodes in the network.

In this paper we introduced a modification to TEEN protocol, which improves the CHs selection. After this modification dead nodes are not selected for cluster heads and all nodes in the network are classified to cluster areas with a use of the K-nearest neighbour algorithm on the basis of the best CHs position to minimize energy consumption and maximize life time of the network. Simulation results performed in MATLAB showed that in modified protocol all nodes are alive, in contrast to standard protocol, where 10 dead nodes were selected for cluster heads, what will cause short life time of the network.

Acknowledgements. This work was financed by the Lodz University of Technology, Faculty of Electrical, Electronic, Computer and Control Engineering as a part of the statutory activity (project no. 501/12-24-1-5438).

References

1. Majeswar, A., Agrawal, D.P.: TEEN: a protocol for enhanced efficiency in wireless sensor networks. In: Proceedings of 1st International Workshop on Parallel and Distributed Computing Issues in Wireless Networks and Mobile Computing, San Francisco, CA, USA, p. 189 (2001)
2. Xu, J., Jin, N., Lou, X., Peng, T., Zhou, Q., Chen, Y.: Improvement of LEACH protocol for WSN. In: 9th International Conference on Fuzzy Systems and Knowledge Discovery, China (2012)
3. Yang, Z., Mohammed, A.: A survey of routing protocols of wireless sensor networks. Blekinge Institute of Technology, Sweden, White Paper (2010)

4. Manjeshwar, A., Agrawal, D.P.: TEEN: a routing protocol for enhanced efficiency in wireless sensor networks. Center for Distributed and Mobile Computing, ECECS Department, University of Cincinnati, Cincinnati, OH 45221-0030
5. Ibrahim, A.A., Tamer, A.K., Abdelshakour, A.: SEC-TEEN: a secure routing protocol for enhanced efficiency in wireless sensor networks (2013)
6. Yong, Z., Youwen, L., Shixiong, X.: An improved KNN text classification algorithm based on clustering. J. Comput. 4(3), 230–237 (2009)
7. Teknomo Kardi: K-Nearest Neighbors Tutorial (2018). http://people.revoledu.com/kardi/tutorial/KNN/index.htm
8. Juwaied, A., Jackowska-Strumiłło, L., Sierszeń, A.: Improved clustering algorithm of LEACH Protocol in IoT Network. In: International Interdisciplinary PhD Workshop, Lodz, Poland, pp. 170–175 (2017)
9. Dahiya, S., Kumar, S.: Modified TEEN protocol for enhancing the life time of sensor network. Int. J. Technol. Res. Eng. 3, 2901–2904 (2016)
10. Juwaied, A., Jackowska-Strumiłło, L.: Analysis of cluster heads positions in stable election protocol for Wireless Sensor Network. In: International Interdisciplinary PhD Workshop (IIPhDW), Swinoujscie, Poland, pp. 367–370. IEEE Xplore (2018)
11. Sierszeń, A., Sturgulewski, Ł., Ciazyński, K.: User positioning system for mobile devices. In: Federated Conference on Computer Science and Information Systems (FedCSIS), pp. 1327–1330. IEEE Xplore (2013)
12. Sierszeń, A., Sturgulewski, Ł., Kotowicz, A.: Tracking the node path in wireless ad-hoc network. In: Federated Conference on Computer Science and Information Systems (FedCSIS), pp. 1321–1325. IEEE Xplore (2013)

The Criteria for IoT Architecture Development

Łukasz Apiecionek(✉)

Kazimierz Wielki University, ul. Chodkiewicza 30, 85-064 Bydgoszcz, Poland
`lapiecionek@ukw.edu.pl`

Abstract. Computer systems are presently a common element of every-day life. In this field the main idea is Internet of Things, which consists in connecting all possible devices to the Internet in order to provide them with new functionalities and new services. There are lot of known types of architecture for the Internet of Things solutions. Of course, each of them provides different opportunities for their users. There is a problem which one will be the best for specific solution. That is why the author propose in this paper the criteria which let to decide which IoT architecture will suite the requirements.

1 Introduction

Computer systems are presently a common element of everyday life. In this field the main idea is Internet of Things, which consists in connecting all possible devices to the Internet in order to provide them with new functionalities and new services [1]. Its objective is to ensure everybodys access to any desired service, at any place and by any possible transmission medium. Presently the Internet of Things can be described as a solution in implementation phase. New architectures emerge all the time but the technology allows developing new services right now. One of the presently available and widely used services is monitoring all kinds of physical values, production lines or processes which accompany them. The solutions provided allow making better decisions. But how to choose the proper IoT architecture? The author made a research on common architecture and propose the criteria which let to choose the network administrator the best solutions for their requirements. Thereby, this paper presents the main network architectures. Section 2 of the article covers the history of the IoT. Section 3 summarizes three basic types of architecture for the solutions. In Sect. 4 presents the criteria which should be considered by network administrator during development of IoT solution. Section 5 concludes the paper.

2 History of IoT State of the Art

The IoT enables physical objects to share information and to coordinate decisions. The IoT changes traditional objects into the so-called smart objects. It is

© Springer Nature Switzerland AG 2019
M. Choraś and R. S. Choraś (Eds.): IP&C 2018, AISC 892, pp. 169–176, 2019.
https://doi.org/10.1007/978-3-030-03658-4_20

achieved by equipping them with sensors, transmission protocols and software in order to allow them to process data and communicate with other devices.

Figure 1 illustrates the overall concept of the IoT, in which every domain specific application is interacting with domain independent services, whereas in each domain, sensors and actuators communicate directly with each other [4].

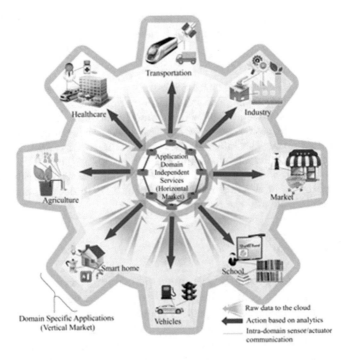

Fig. 1. The overall picture of IoT emphasizing the vertical markets and the horizontal integration between them [4]

It is expected that in a course of time much more devices will be connected to the Internet and altogether they will create an intelligent environment. Synchronizing the solutions will allow for example an earlier opening of the garage door when the car approaches the premises. Using the intelligent systems of transport will enable more efficient traffic control, including preventing congestions or ensuring the emergency vehicles right-of-way by manipulating traffic lights. However, it requires overcoming numerous obstacles. Frequent problems in this matter include unfortunately:

- the necessity of providing power supply for all elements in the IoT solution;
- the necessity of transmitting the data to remote destinations within the area the solution covers;
- the necessity of developing data transmission protocols how to send the data, how to transmit them efficiently;

- creating a datacenter for collecting and processing the data where and how to store big amounts of data, how to share them;
- developing the algorithms for data analysis how to analyze the data, how to draw adequate conclusions;
- the necessity of connecting various devices which may have been incompatible before how to connect various elements, what converters and what gateways are to be developed;
- addressing the data how to address and identify the devices.

3 Common IoT Architecture

In the pursue of the universal architecture the existing ones should not be disregarded. In the course of analysis of the present-day offers, they may be divided in regards to the place of data processing: in the sensor itself or in a central point. Thus, the following types of architectures can be specified:

- Sensors,
- Fog,
- Cloud Computing.

The above solutions are illustrated in Fig. 2. At the lowest level there are sensors the IoT devices which are mainly responsible for collecting data. They are and will be the most numerous ones, as they are accountable for connecting the solutions to the network. Due to their great number, they generate the highest requirements for the address pools and the main network traffic. The upper layer is the Fog-type solution, which gathers, aggregates and preliminarily processes the collected data. Such approach fosters lowering the network traffic generated by the sensors. It is estimated that millions of IoT sensors would generate a lot of unnecessary traffic, and Fog layer could help to deal with this problem. The equipment used in Fog layer possess a higher computing power and in most cases they require different working conditions. The upmost layer is Cloud. In this type of solution all data are processed in the cloud. It requires a suitable structure, building a center for data processing and managing it in a proper manner, but also sending all the data to the center, where their processing takes place.

All layers may exist separately or in connection with others, providing services for the other ones. The Sensors layer covers not only the sensors themselves, which measure physical values, but also the actuators, responsible for controlling, changing and setting of physical values. This layer requires developing adequate converters, which may be active, passive, mechanical, optical, magnetic, thermal, electrical, biological, chemical etc. Fog layer could also work as a bridge between the Sensors and Cloud level, aggregating the collected data and lowering the network traffic. It allows to move the data processing services closer to the terminal elements' the sensors. Fog layer also enables earlier reaction to the data obtained from the sensors.

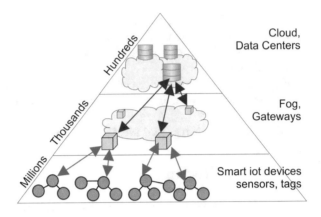

Fig. 2. The IoT elements [4]

It is recommended to the designers due to the following features [4]:

- Location: Fog resources are positioned between smart objects and the cloud data centers; and thus, providing better delay performance.
- Distribution: Since fog computing is based on micro centers with limited storage, processing and communication capabilities in comparison to the cloud, it is possible to deploy many such micro centers closer to the end users, as their cost is usually a small fraction when compared to cloud data centers.
- Scalability: Fog allows IoT systems to be more scalable, i.e. when the number of end users increases, the number of deployed micro fog centers can follow, in order to cope with the growing load. Such increase cannot be achieved by the cloud because the deployment of new data centers is cost-prohibitive.
- Density of devices: Fog helps to provide resilient and replicated services.
- Mobility support: Fog resources act as a mobile cloud, as it is located close to the end users.
- Real-time: Fog has the potential to provide better performance for real-time interactive services.
- Standardization: Fog resources can interoperate with various cloud providers.
- On-the-fly analysis: Fog resources can perform data aggregation in order to send partially processed data, instead of raw data, to the cloud data centers for further processing.

Fog solutions can be deployed hierarchically, in result of which they compose multi-tier solutions, as it is demonstrated in Fig. 3.

Employing CC for the IoT involves the following challenges [4]:

- Synchronization: Synchronization between different cloud providers poses a challenge to offer real-time services, since they are built on top of various cloud platforms.
- Standardization: Standardizing CC is also a significant issue for the IoT cloud-based services, due to a necessity to interoperate with various providers [7,8].

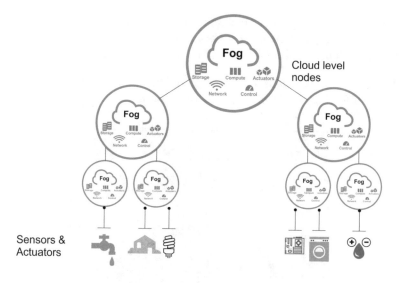

Fig. 3. Multi-tier deployment [3]

- Balancing: achieving balance between general cloud service environments and the IoT requirements may raise difficulties due to the differences in the infrastructure.
- Reliability: The security of the IoT cloud-based services presents another challenge due to the differences in the security mechanisms between the IoT devices and the cloud platforms [2,5].
- Management: Managing CC and the IoT systems is also a demanding task, as they use different resources and components.
- Enhancement: Validating the IoT cloud-based services is necessary to ensure providing high-quality services that meet the customers expectations [6],
- Network Transfer: collecting all data in CC generates an intense data traffic through the network equipment.

4   The Criteria for Architecture Scheme Development

Selecting the architecture for the solution could depends on numerous factors. In order to choose the most optimal one, the analysis of the following issues should be performed:

- what means of communication are to be applied in the solution, what is the transmission capability of the designed solution, is there any already existing network architecture or should it be set up from a scratch;
- what is the bit error rate in the applied network;
- what is the amount of the data which are to be transmitted by the solution;
- what is the number of the data receivers;
- how many sensors are to be applied,

– what is the desired memory capability of the applied sensor,
– what power supply is planned for the solution,
– what is the longest admissible time of failure of the elements in the solutions.

The above considerations are well depicted in Fig. 4, which shows four aspect profoundly affecting the architecture of the solution.

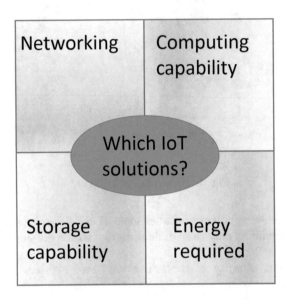

Fig. 4. The IoT architecture criteria

The proposed criteria for the analysis and possible solutions of IoT architecture are presented in Table 1.

Table 1. Criteria for the analysis of the solution.

Criterion	Available options
Means communication	Radio wireless, cable
Bit error rate in the transmission	A percentage rate of incorrect packets in the transmission
Size of the data for the transmission	The number of bytes ti be sent per seconds
The number of data recipients	Several, over a dozen, several dozens, hundreds, thousands
Memory capability of the sensors	kB, MB, GB
Power supply	Battery, mains
Longest admissible time of failure of the elements in the solution	Minutes, hours, days

Choosing the most advantageous solution must be preceded by gathering the answers for all the above questions, though obviously there are also other influential factors, such as budget of the planned solution, which may be conclusive as to whether the solution will cover all three layers, i.e. Sensors, Fog and CC. The selection of routing in the network will be dependent, in turn, on the work format of the solution, whether it will be:

- cable network,
- mobile network,
- a combination of cable and mobile network.

5   Conclusion

The IoT solutions are in implementation phase right now. The IoT systems are meant to provide new services which help to make life easier, accelerate some processes, increase the efficiency of production while lowering its cost at the same time. This paper presents the criteria which should be taken into account during the preliminary analysis on which architecture is the most optimal for solving the given problem, especially as the IoT is dedicated to solve specific problems and provide specific services. Moreover, four factors affecting the architecture were taken into account:

- networking,
- computing capability,
- storage capability,
- energy required.

It is the fact of possessing or lacking the power supply in the Sensors layer which is conclusive about many features of the system under construction, i.e. which sensors to select, how to design them or which means of communication can be used. As a result of the performed analysis, the criteria for choosing the best network architecture were proposed.

Acknowledgment. This article was funded under the grant Miniatura 1 number 2017/01/X/ST6/00613 from National Science Centre, Poland.

References

1. Al-Fuqaha, A.I., Aledhari, M., Ayyash, M., Guizani, M., Mohammadi, M.: Internet of Things: a survey on enabling technologies, protocols, and applications. IEEE Commun. Surv. Tutor. **17**, 2347–2376 (2015)
2. Grobmann, M., Illig, S., Matejka, C.: Environmental monitoring of libraries with MonTreAL. In: Kamps, J., Tsakonas, G., Manolopoulos, Y., Iliadis, L., Karydis, I. (eds.) Research and Advanced Technology for Digital Libraries, TPDL 2017. Lecture Notes in Computer Science, vol. 10450. Springer, Cham (2017). https://doi.org/10.1007/978-3-319-67008-9_52

3. Harald Sundmaeker, P.F., Guillemin, P., Woelffl, S.: Vision and challenges for realising the Internet of Things. Pub. Office EU (2010). http://www.internet-of-thingsresearch.eu/pdf/IoT_Clusterbook_March_2010.pdf
4. Khan, R., Khan, S.U., Zaheer, R., Khan, S.: Future internet: the Internet of Things architecture, possible applications and key challenges. In: 2012 10th International Conference on Frontiers of Information Technology (FIT), pp. 257–260 (2012)
5. Kozik, R., Choras, M., Ficco, M., Palmieri, F.: A scalable distributed machine learning approach for attack detection in edge computing environments. J. Parallel Distrib. Comput. **119**, 18–26 (2018)
6. Kozik, R., Choras, M., Holubowicz, W.: Packets tokenization methods for web layer cyber security. Log. J. IGPL **25**(1), 103–113 (2017)
7. Makowski, W., Motylewski, R., Stosik, P., Jelinski, M., Apiecionek, L.: Jakosc i wydajnosc transmisji danych w sieci Wi-Fi MESH WDS systemu monitorowania urzadzen eksploatowanych w strazy pozarnej, Wspomaganie procesow zarzadzania dzialaniami w strazy pozarnej/red. Jacek Rogusk, str. 39–47, Jozefow (2016)
8. Wu, M., Lu, T.J., Ling, F.Y., Sun, J., Du, H.Y.: Research on the architecture of Internet of Things. In: 2010 3rd International Conference on Advanced Computer Theory and Engineering (ICACTE), pp. 484–487 (2010)

The Problem of Sharing IoT Services Within One Smart Home Infrastructure

Piotr Lech[(✉)]

Department of Signal Processing and Multimedia Engineering, Faculty of Electrical Engineering, West Pomeranian University of Technology, Szczecin, 26 Kwietnia 10, 71-126 Szczecin, Poland
piotr.lech@zut.edu.pl

Abstract. The article presents the risk of sharing the communication link for IoT (Internet of Things) services. The integration of two popular smart home systems - video surveillance system and home automation - has been considered. The case in which data processing, analysis and development of control is carried out on an external server is analysed. The computer simulation has used services that unevenly and asymmetrically loaded the bandwidth. In the study, both the link state and RESTful failures were reported simultaneously.

Keywords: IoT · Video surveillance · Services stability

1 Introduction

The recently noticeable dynamic development of technologies used in the Internet of Things systems raises many problems related to ensuring failure-free operation of them. The recent migration of control applications in smart home systems from local resources to the cloud exposes such solutions to failures. In particular, the transfer of control algorithms in IoT systems from the local domain to the cloud carries the risk of losing controllability of these objects. In this paper, it has been shown how dangerous might be the use of highly link loading services together with the others which use the link to a lesser extent. In particular, applications that use SOAP (Simple Object Access Protocol), REST (Representational State Transfer), etc., are vulnerable to failures [3,6,8]. Expecting that successful calls using these technologies will always work is a utopia. Unfortunately, many programmers treat the Web API (WebServices Application Programming Interface) remote calls in the same way as local calls. This approach, in combination with the heavily-charged service link, most often leads to error handling caused by disconnection. An example of this type of problems can be simultaneous use of smart home observation sets together with control of IoT devices (Fig. 1). When configuring video transmission parameters of observation sets, it is easy to make a mistake resulting in overloading of the link. In addition, it will affect the stability of the application controlling all devices in a smart home.

© Springer Nature Switzerland AG 2019
M. Choraś and R. S. Choraś (Eds.): IP&C 2018, AISC 892, pp. 177–184, 2019.
https://doi.org/10.1007/978-3-030-03658-4_21

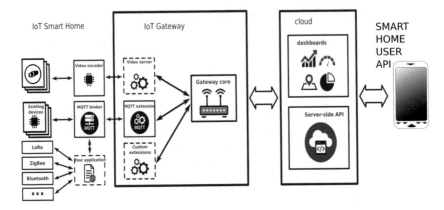

Fig. 1. Example of the IoT smart home system

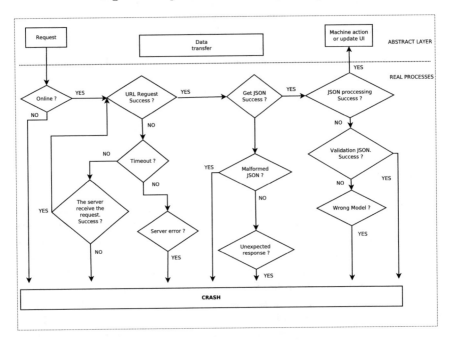

Fig. 2. RESTful API potential way to crash

Analysing potential errors during the calls for RESTful APIs network requests, illustrated in Fig. 2, the development of a reliable application using this technology can be considered as a challenging task. There may be many sources of failures (Fig. 2). In the analysed case of the coexistence of two IoT systems the reason of the failure is the instantaneous loss of communication resulting from the link overload. Looking at the most common failures ranked in order from the least frequent to frequently occurring, the following cases can be distinguished:

- **failures of DNS services.** - a very rare event, but effectively blocking the establishment of transmission,
- **data center failures** - failures of servers operating in the cloud responsible for the SaaS (Software as a Service) or PaaS (Platform as a Service) service are rare, for example, Amazon's AWS service has an availability guarantee of 99.95% – such good results are encouraged to create systems that use Web APIs,
- **loss of validity of SSL certificates** - a failure rarely found on the server side, but not avoidable on the client's side,
- **communication errors** - they generate failures due to various reasons, starting from physical damages of the transmission medium, through transmission configuration errors, ending with improper implementation of communication standards.

Not only physical connection loss can be a source of loss of stability of the application – some other reasons can be e.g.: syntax errors, improper redirects, credential's expiration or failures of proxy servers.

2   Smart Services with One-Link Sharing

A dangerous threat to the stability of IoT systems when sharing the same link is its peak overload [5]. This overload may take a few seconds (or even minutes) which may lead to the connection breaking. In this case, the transmission of the video stream is most often broken (usually in this case the transmission is repeated automatically after re-establishing the connection) [4]. Applications using the Web API can be configured in such a way that the connection with the client does not end, but most often in such a situation the connection is deleted by the firewall due to too long inactivity.

Most households have one Internet connection with a fixed maximum bandwidth. In this case, home video surveillance systems used in property protection systems share the connection with home automation devices. Data from local network has been sent to applications in the cloud using the RESTful API. A large disproportion of the available bandwidth is noticeable (Fig. 3). The assumption was made that users do not charge a common link by carrying out private activities (watching video, listening to music, browsing websites). The above simplification makes sense when the residents are away from home.

Emerging failures may lead to unexpected problems, e.g. in the case of video observation, we may overlook a robbery or fire breakdown as a result of the connection loss, and in the home automation system we can expect the device to be switched on but as a result of a transmission failure this does not occur.

The cause of the failure in the form of link overloading may be erroneous resource reservation resulting from improper parameterization of video streams. Even the choice of CBR (Constant Bit Rate) or VBR (Variable Bit Rate)

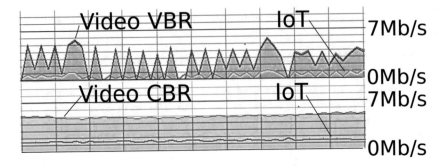

Fig. 3. Example of traffic for Video and IoT data transfer (with 7 Mb/s bandwidth limitation and proper data stream configuration)

transmission leads to certain consequences. Reservation of bandwidth for CBR transmission in the case of leaving too little margin for transmission associated with home automation will generate a failure, and when using the QoS (Quality of Service) for video transmission, it will be a failure of home automation systems. In the case of overestimating the maximum load capacity in VBR mode, it may lead, depending if the QoS is used or not, to a total failure or affecting only the home automation system. The use of multi-camera systems additionally increases the probability of link saturation and the generation of an emergency state.

3   Experimental Tests Problem in Real Smart Home Network Environment

The case described in this paper is a kind of tests being difficult to automate due to various configuration requirements of cooperating systems. In the above considerations, we have two systems prone to failure: one with a large transfer cost and the second with a negligible one. Even the adopted simplification regarding the lack of traffic generated by users in the area other than IoT-related traffic does not allow correct analysis of the phenomena occurring in the system due to the high variability of available Internet bandwidth [1].

The solution of this problem might be the simulation carried out in a fully predictable environment including networks and devices.

4   Simulation

One of the methods of creating fail-safe applications is the verification of their correctness by simulating unexpected events. Testing for failures is the best way to prevent future unpleasant surprises. However, not all tests can be created automatically.

The configuration diagram of virtual machines based on VirtualBox [2] and their connections used in the simulation is presented on Fig. 4. For testing purposes the RESTful server and client applications have been created in Python 3 due to low resource utilization which is extremely important in virtualized systems [7].

To understand the phenomenon of mutual interaction of systems, some simplifications have been introduced:

- a single video stream from VLC player (VideoLan media player) is used,
- the metering and control data stream is constant over time and is much smaller than the video stream,
- control of home appliances (along with the collection of data from sensors generated on VM2 virtual system in Fig. 4) developed by an external web application controlling a smart home (VM1 - Fig. 4),
- the video stream is hosted by VLC server on the same machine as the IoT system (VM2 - Fig. 4),
- the bandwidth has been controlled by the VM3 router (Fig. 4)
- all of the human interactions with IoT system have been emulated on VM4 in Fig. 4).

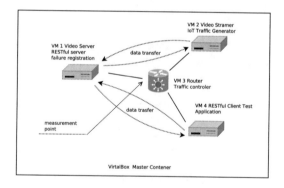

Fig. 4. VirtualBox and Linux powered simulation environment

All transmission failures along with the transmission parameters at which the communication has been interrupted are recorded.

The study was conducted according to the following scenarios:

- CBR video transmission and control data stream without QoS
- CBR video transmission and control data stream with defined QoS
- VBR video transmission and control data stream without QoS
- VBR video transmission and control data stream with defined QoS

In addition, two states have been checked in all cases:

- the attachment of the link via the video stream is close to the limit of the maximum bandwidth and the addition of the transmission associated with the control data results in exceeding this limit (this value is changed step by step),
- the attachment of the link via the video stream is close to the maximum bandwidth limit and the addition of the transmission associated with the control data does not exceed this limit.

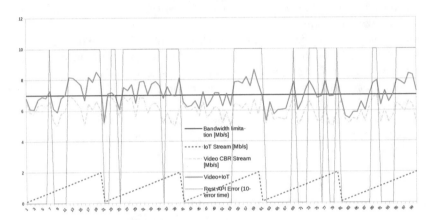

Fig. 5. Video CBR without QoS, IoT data stream, RESTful API errors, 7 MB/s bandwidth limitation, measurement interval delay 10 ms

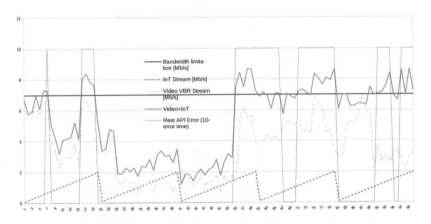

Fig. 6. Video VBR without QoS, IoT data stream, RESTful API errors, 7 MB/s bandwidth limitation, measurement interval delay 10 ms

Selected test results without QoS are presented in Figs. 3, 5 and 6. It is clearly visible that there is such a configuration both in VBR and CBR modes for which there are no failures. However, system parameterization for VBR transmissions for inexperienced administrators is problematic due to the need of observing and recording the maximum data transfer rates of the link.

In the case when the QoS is used, transmission of measurement and control data always fails. In other cases, emergency states are randomly distributed between subsystems.

Because the web application crashes have been recorded at the same time, an interesting relationship has been observed between the delay and the occurrence of the link failure, and the fall of the application. On one hand, this may result from emptying the transmission buffers and on the other hand from the delay after which the server side application detects a transmission failure. Typical TCP transmission failures occur within a few dozen ms, but in the case of REST applications for keeping the response, much longer times of even a few seconds are used. In the case of information websites, it makes no sense as the maximum response times expected by users in the IoT systems are about 100–200 ms.

It should be noted that in order to observe the global behavior of the system, in the prepared simulation experiments, it was taken into account that the system response should not last longer than the interval between measurements. This suggestion has been included in the configuration of server parameters for the maximum time of handling requests and lack of connection persistence.

5   Conclusions and Future Work

In the above considerations the issue of loss of connection for a video signal has been omitted, assuming the view of the frozen picture frame as satisfactory. In such case the higher priority of intelligent home control above the video surveillance has been assumed.

Integration of various IoT services related to data processing which implement business logic on external servers is currently an irreversible trend. Despite the excellent results related to the reliability of data centers, there is a risk of system stability loss due to bad, unmatched transmission conditions through a shared link. It has been shown, on the basis of the simulation, that for asymmetric transmission (in terms of band occupancy), it can lead to loss of stability if the service is poorly parameterized. At the same time, some connections between the occurrence of a link failure and the failure of the application have been noticed. Probably, in some cases, a properly configured firewall configuration with a proxy server should prevent this type of failure, however additional tests are required to confirm this observation. The recommendation to counteract the failure associated with the overloading of the link is to comply with the requirement to test the system under overload conditions and appropriate configuration of services.

References

1. Amar, Y., Haddadi, H., Mortier, R., Brown, A., Colley, J., Crabtree, A.: An analysis of home IoT network traffic and behaviour. arXiv preprint arXiv:1803.05368 (2018)
2. Arora, D.S., Majumder, A.: Methods, systems, and computer readable media for packet monitoring in a virtual environment. US Patent 9,967,165 (2018)
3. Kim, S., Hong, J.Y., Kim, S., Kim, S.H., Kim, J.H., Chun, J.: Restful design and implementation of smart appliances for smart home. In: IEEE 11th International Conference on Ubiquitous Intelligence & Computing and IEEE 11th International Conference on Autonomic & Trusted Computing and IEEE 14th International Conference on Scalable Computing and Communications and Its Associated Workshops (UIC-ATC-ScalCom), pp. 717–722. IEEE (2014)
4. Laghari, A.A., He, H., Channa, M.I.: Measuring effect of packet reordering on quality of experience (QoE) in video streaming. 3D Res. **9**(3), 30 (2018)
5. Lech, P., Włodarski, P.: Analysis of the IoT WiFi mesh network. In: Silhavy, R., Senkerik, R., Kominkova Oplatkova, Z., Prokopova, Z., Silhavy, P. (eds.) Cybernetics and Mathematics Applications in Intelligent Systems, pp. 272–280. Springer (2017)
6. Petrillo, F., Merle, P., Moha, N., Guéhéneuc, Y.G.: Are REST APIs for cloud computing well-designed? An exploratory study. In: International Conference on Service-Oriented Computing, pp. 157–170. Springer (2016)
7. Richardson, L., Ruby, S.: RESTful Web Services. O'Reilly Media Inc., Sebastopol (2008)
8. Sanchez, B.A., Barmpis, K., Neubauer, P., Paige, R.F., Kolovos, D.S.: RestMule: enabling resilient clients for remote APIs (2018)

Monitoring - The IoT Monitoring Tool for Fire Brigades

Łukasz Apiecionek[1(✉)] and Udo Krieger[2]

[1] Kazimierz Wielki University, ul. Chodkiewicza 30, 85 064 Bydgoszcz, Poland
lapiecionek@ukw.edu.pl
[2] University of Bamberg, Bamberg, Germany
udo.krieger@uni-bamberg.de

Abstract. In the field of computer systems the world has entered an era of Internet of Things, which consists in connecting all possible devices to the Internet in order to provide them with new functionalities and in this way to improve the user life standard. There are lot of systems which required the possibility for monitoring their condition. One of the already implemented solution is Monitoring the system prepared for fire brigade dedicated for monitoring their equipment. In this paper the authors presented the architecture of fire brigade monitoring tool.

1 Introduction

The Internet of Things idea consists in connecting all possible devices to the Internet in order to provide them with new functionalities and in this way to improve the user life standard [1]. It has to ensure everybody access to any desired service, at any place and by any possible transmission medium. Presently the Internet of Things can be described as a solution in progress. New architectures emerge all the time and at the same moment the technology allows developing new services [2,3]. One of the presently available and widely used services is monitoring all kinds of physical values, production lines or processes which accompany them. One of the already implemented system is Monitoring which is described in the Sect. 3. But first, Sect. 2 of the article covers the history of the IoT.

2 History of IoT State of the Art

As it was mentioned, the IoT enables physical objects to share information and to coordinate decisions. The IoT changes traditional objects into the so-called smart objects. It is achieved by equipping them with sensors, transmission protocols and special software. All of them together allow to process data and communicate with other devices. Fig. 1 illustrates the overall concept of the IoT, in which every domain specific application is interacting with domain independent services, whereas in each domain, sensors and actuators communicate directly with each other [4].

© Springer Nature Switzerland AG 2019
M. Choraś and R. S. Choraś (Eds.): IP&C 2018, AISC 892, pp. 185–191, 2019.
https://doi.org/10.1007/978-3-030-03658-4_22

Fig. 1. The overall picture of IoT emphasizing the vertical markets and the horizontal integration between them [4]

Connecting devices generates also some problems. The problems in this matter mainly include:

- the necessity of providing power supply for all elements in the IoT solution;
- the necessity of transmitting the data to remote destinations;
- the necessity of developing data transmission protocols;
- creating a datacenter for collecting and processing;
- developing the new algorithms for big data analysis;
- the necessity of providing gateways for connecting different devices;
- addressing schema.

Lot of systems has to be created and developed by some research [5]. One of such solutions is Monitoring.

3 Monitoring Fire Brigade Monitoring Tool

The monitoring system is an IoT solution for monitoring the equipment of fire brigades. The general idea behind the system is presented in Fig. 2. This systems cover all of the general architecture layer for IoT systems: sensors layer, fog layer, cloud layer and middleware layer.

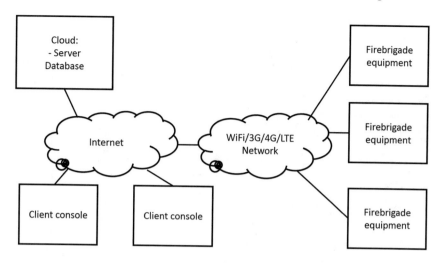

Fig. 2. The idea of the monitoring system

Table 1. The elements of the analyzed IoT system

IoT Elements		Monitoring system
Identification	Naming	DNS
	Addressing	IPv4
Sensing		Smart sensors, GPS
Communication		WiFi, 3G/4G/LTE
Computation	Hardware	Proprietary project with ARM processor
	Software	Linux
Service		Data analysis, failure prediction
Semantic		–

The analysis of the Monitoring solution is summarized in Table 2, which lists the elements, protocols, and standards implemented in the system.

In the Sensors layer of IoT systems the following devices are connected: pumps, electric shears and water tender. The range of their monitoring covers also their working parameters, which are listed, along with the types of the devices, in Table 1.

The next layer of the systems are the data acquisition devices (presented in Fig. 3) responsible for collecting the data from the sensors and forwarding them, as well as collecting the locations of the devices using GPS, which allows to locate them. Those devices can be regarded as Fog layer according to IoT general architecture schema.

The system is equipped with central solution in type of Cloud, which stores the data collected from the Fog layer. For the purpose of data transmission the architecture was designed as depicted in Fig. 4. Collecting the data from the

188 L. Apiecionek and U. Krieger

Table 2. The parameters of the devices under the system monitoring

	Equipment					
	Niagara 1	Honda SST50	Tohatsu VC72 AS	Autpomp A24/8	Autopomp A16/8	Hydraulic scissors
Fuel flow	x	x	x			
Fuel level	x	x	x			
Discharge pressure	x	x	x	x	x	
Suction pressure		x	x	x	x	
Water flow meter	x	x	x	x	x	
Oil temperature	x	x	x			x
Exhaust temperature	x	x	x			
Water temperature - suction	x	x	x	x	x	
Water temperature - pressing	x	x	x	x	x	
Throttle position	x	x	x			
Engine/shaft revolutions	x	x	x	x	x	
Pressing pressure high stage				x	x	
Engine temperature						x
Oil pressure						x
Power supply voltage						x
Power consumption						x

sensors was performed using cable connections. In the Fog layer for the purpose of connection a WiFi mesh network was selected. Transmission to the Cloud layer was made via a 3/4G network.

Setting a mesh network when all devices are connected in a hop-by-hop manner results of course in a decrease in transmission speed, which was nevertheless sufficient for this network operation. The drop in the transmission speed dependently on the number of hop-by-hop connections is presented in Table 3.

The Middleware layer was constructed using WiFi client connection for mobile tablets, used by the firefighters during the action or using the Internet from any place. For this purpose, two versions of the application were built a client mobile version for tablets and a desktop version for the management center, working near the Cloud layer (as shown in Fig. 5). In the Cloud layer the collected data serve as a basis for the analysis and concluding which piece of equipment is the most susceptible to damage in the nearest future.

The security issues of the system include data encryption, unauthorized use and the DoS-type attacks on the system [7,8]. It is still in the testing phase and those problems are being solved now.

The system allows to monitor the equipment in use. For instance, the process of pumping out water from a flooded basement takes hours. Until present, a firefighter had to supervise the whole action, but now it is possible to leave the pump under the system monitoring and in case of any problems it will set an alarm.

The architecture of the system was affected by the issue of power supply. All the devices possess an external power supply unit (for example a generator),

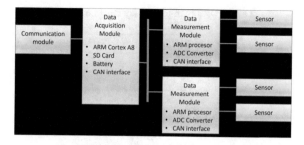

Fig. 3. The architecture of the data acquisition devices in the Fog layer and the sensors in the Sensors layer

but their voltage level is unstable. Thus, it was necessary to construct a suitable battery supply system, which was charged by the power supply, which solved the problem.

The second element which had a profound impact on the architecture of the system was the issue of communication between Fog and Cloud layers. Due to the fact that the already existing communication system was an analogue one, digital communication had to be provided. The cost of building such system was also taken into account and it was designed in the manner which would

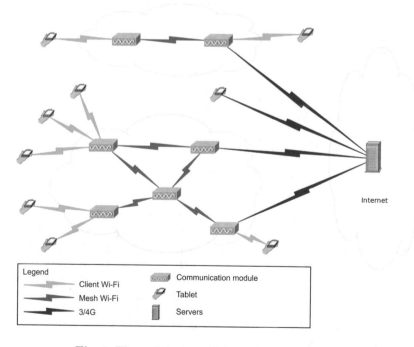

Fig. 4. The architecture of the communication system

Table 3. Throughput by hop count [6]

Hop count	Throughput [MBit/s]
1	272.81
2	141.54
3	69.35
4	35.48
5	18.45
6	8.62

Fig. 5. Screenshots from application software

Table 4. The elements of the analyzed IoT systems

IoT Elements		Monitoring system
Identification	Naming	DNS
	Addressing	IPv4
Sensing		Smart sensors, GPS
Communications		WiFi, 3G/4G/LTE
Computation	Hardware	Proprietary project with ARM processor
	Software	Linux
Service		Data analysis, failure prediction, equipment position monitoring
Semantic		–

make it most cost-effective in use. The amount of data sent by the system was calculated. A device taking 15 measurements with one-second sampling utilizes 7 MB of data per hour. On this basis it was decided to use a 3G/4G network with a pre-paid card, allowing for the transfer of 300 MB of data per month in the 4G LTE standard and after exceeding it with 32 kbps bit rate. Such transfer speed enables to send 14 MB of data during one hour of work, which is a sufficient value. The cost of maintenance of such card between 2017 and 2018 in Poland, where the system was built, was 1.25 Euro per year.

The analysis of the Monitoring solutions is summarized in Table 4, which lists the elements, protocols and standards implemented in the systems.

4   Conclusions

The monitoring system is an IoT solution example for monitoring the equipment of fire brigades. It covers all of the layers from general architecture for IoT systems:

- in the Sensors layer the following devices are connected: pumps, electric shears, and water tender;
- in the Fog layer a WiFi mesh network is used;
- the Cloud layer stores the data collected from the Fog layer;
- the Middleware layer was constructed using WiFi client connection for mobile tablets.

This system is using already existing technologies and open source solutions which allows them achieve quickly new capabilities and launch new services for the users. It was developed and implemented in real fire brigade and is steal working and generating data for future analysis.

Acknowledgement. This article was funded under the grant Miniatura 1 number 2017/01/X/ST6/00613 from National Science Centre, Poland.

References

1. Harald Sundmaeker, P.F., Guillemin, P., Woelfflé, S.: Vision and challenges for realising the Internet of Things. Pub. Office EU (2010). http://www.internet-of-thingsresearch.eu/pdf/IoT_Clusterbook_March_2010.pdf
2. Khan, R., Khan, S.U., Zaheer, R., Khan, S.: Future internet: the Internet of Things architecture, possible applications and key challenges. In: 10th International Conference on Frontiers of Information Technology (FIT), pp. 257–260 (2012)
3. Wu, M., Lu, T.J., Ling, F.Y., Sun, J., Du, H.Y.: Research on the architecture of Internet of Things. In: 3rd International Conference on Advanced Computer Theory and Engineering (ICACTE), pp. 484–487 (2010)
4. Al-Fuqaha, A.I., Aledhari, M., Ayyash, M., Guizani, M., Mohammadi, M.: Internet of Things: a survey on enabling technologies, protocols, and applications. IEEE Commun. Surv. Tutor. **17**, 2347–2376 (2015)
5. Grobmann, M., Illig, S., Matejka, C.: Environmental monitoring of libraries with MonTreAL. In: Kamps, J., Tsakonas, G., Manolopoulos, Y., Iliadis, L., Karydis, I. (eds.) Research and Advanced Technology for Digital Libraries. TPDL 2017. LNCS, vol. 10450. Springer, Cham (2017). https://doi.org/10.1007/978-3-319-67008-9_52
6. Makowski, W., Motylewski, R., Stosik, P., Jelinski, M., Apiecionek, L.: Jakosc i wydajnosc transmisji danych w sieci Wi-Fi MESH WDS systemu monitorowania urzadzen eksploatowanych w strazy pozarnej. In: Rogusk, J. (ed.) Wspomaganie procesow zarzadzania dzialaniami w strazy pozarnej, Jozefow, pp. 39–47 (2016)
7. Kozik, R., Choras, M., Ficco, M., Palmieri, F.: A scalable distributed machine learning approach for attack detection in edge computing environments. J. Parallel Distrib. Comput. **119**, 18–26 (2018)
8. Kozik, R., Choras, M., Holubowicz, W.: Packets tokenization methods for web layer cyber security. Log. J. IGPL **25**(1), 103–113 (2017)

Sparse Autoencoders for Unsupervised Netflow Data Classification

Rafał Kozik[(✉)], Marek Pawlicki, and Michał Choraś

UTP University of Science and Technology in Bydgoszcz, Bydgoszcz, Poland
`rafal.kozik@utp.edu.pl`

Abstract. The ongoing growth in the complexity of malicious software has rendered the long-established solutions for cyber attack detection inadequate. Specifically, at any time novel malware emerges, the conventional security systems prove inept until the signatures are brought up to date. Moreover, the bulk of machine-learning based solutions rely on supervised training, which generally leads to an added burden for the admin to label the network traffic and to re-train the system periodically. Consequently, the major contribution of this paper is an outline of an unsupervised machine learning approach to cybersecurity, in particular, a proposal to use sparse autoencoders to detect the malicious behaviour of hosts in the network. We put forward a means of botnet detection through the analysis of data in the form of Netflows for a use case.

1   Introduction

Network and information security is presently one of the most pressing problems of the economy, a major concern for the citizens, a serious matter for the contemporary society and a crucial responsibility for homeland security. However, the immense growth of Internet users generated a plethora of adversaries who abuse the Internet's framework. Currently, the number of successful attacks on information, citizens, and even secure financial systems is still growing. Some attacks are performed by malicious users acting alone, some are carefully arranged invasions performed by groups of compromised machines. In this paper, as a use case scenario, we consider the problem of botnet detection by means of analysing the data in form of NetFlows. The problem of botnets is related to the situation where massive numbers of computers have been infected, through an array of methods, like e-mail attachments, drive-by downloads etc. The infected machines from a kind of network controlled by a botmaster, who issues a fire signal to cause malicious activities. The problem of botnets is highly relevant, as these can be responsible for DoS attacks, spam, sharing or stealing data, fraudulent clicks and many other. NetFlow, often abbreviated to simply flow is a derivative of a data stream shared between two systems. It records comprise a statistic of traffic between the same IP addresses, same source and destination ports, IP protocols and IP Types of service. The reason, why researchers invest efforts to develop the mechanism for NetFlow-based detection techniques is of two-fold. Firstly, the

© Springer Nature Switzerland AG 2019
M. Choraś and R. S. Choraś (Eds.): IP&C 2018, AISC 892, pp. 192–199, 2019.
https://doi.org/10.1007/978-3-030-03658-4_23

nature of the NetFlow data structure allows overcoming the problems related to the privacy and sensitiveness of information that underlying data can exhibit [3]. Secondly, NetFlow standard is widely used by the network administrators as a useful tool for a network traffic inspection purposes.

The remainder of the paper is organized as follows: in Sect. 2 the state of the art in network anomaly detection is given. Section 3 contains the description of the autoencoder-based unsupervised approach to cybersecurity, whereas in Sect. 4 the results obtained on malware datasets are presented and discussed. The paper is concluded with final remarks and plans for future work.

2   Related Work

One of the challenges of combating the botnets is to identify the botmaster in order to block the communication with the infected machines. Currently, the malware is using Domain Generation Algorithms (DGA). DGAs are a way for botnets to hide the Command and Control (C&C) botmaster server. When the botmaster server is compromised, it loses command over the entire botnet. Therefore, anti-virus companies and OS vendors blacklist its IP and stop any possible communication at the firewall level. In [2] a NetFlow/IPFIX based DGA-performing malware detector is suggested. DGA malware is expected to attempt to try to contact more domains than it does new IP addresses. Because of the NetFlows unidirectionalness, additional information has to be used to single out the originator of each transmission. NetFlows with the same IP-port-protocol triples are paired and marked as request and response according to timestamps since a request always comes first. However, this method loses its reliability when scaled to larger networks. In such cases, a service detection algorithm provides a strong feature based on a median number of peers difference. A DNS anomaly detector is implemented, labeling the right tail of the normal distribution as anomalous. This is because DNS requests that are more numerous than the visited IPs are possible C&C botnet connections. Anomaly values are acquired with a fuzzy function. Finally, as the proposed method is susceptible to raising a false positive for DNS resolver service, data from the service detection step is used to tackle this problem.

A real-time intrusion detection system (IDC) with a hardware-core of High-Frequency Field Programmable Gate Arrays was presented in [4]. The authors examine a batch of IDC's including the recent additions to the field, which are based on computational intelligence, including fuzzy systems, support vector machines, and evolutionary algorithms like differential evolution, genetic algorithms or particle swarm optimisation. Whereas these software-based methods are able to quickly adapt to new threats, their effectiveness in high-volume environments is limited by their detection speed. This translates into the inability to properly address the needs of massive environments in the near future, like cloud computing. The proposed hardware-cored IDS offers a higher detection speed than the software-based counterparts. The method supplies a Field-Programmable Gate Array (FPGA) with an internally evolvable Block Based

Neural Network (BBNN). The BBNN in this process is a feed-forward algorithm. The character of the blocks and the weights are determined by a genetic algorithm which seeks a global optimum guided by a specific fitness function. NetFlow data is used as a way to streamline feature extraction, as these can be set to default flow features. Additionally, the NetFlow collector is able to generate real-time data for the FPGA. The procedure itself is as follows: the FPGA performs real-time detection of possible intrusions and adds the record to the database; the BBNN repeatedly re-trains itself with the fresh database; the FPGA corrects its configuration building on the structure of the BBNN.

In [5] authors proposed the use of a micro-cluster based outlier detection algorithm (MCOD) as a deviation discovery device, augmented to take into account pattern deviation over time. The procedure employs clustering to cut down on the number of distance calculations it has to perform. The reduced calculation needs make it suitable for real-time data stream analysis, unlike many other anomaly detection approaches. The distance between a flow and a cluster centroid decides whether it is an anomaly, or not. MCOD is used in a succession of intervals, making the algorithm time-aware. The effects of the anomaly detection are then processed by a polynomial regression to arrive at an approximation of cluster densities over time. In the proposed procedure, all cluster densities are monitored disregarding the cluster edge denoted by the k variable. This approach diminishes the impact of the unlikely supposition that all traffic is distributed equally across the network, allowing for an increased situational awareness. By comparing cluster densities over time two polynomials are generated to represent the cluster activity. Overall, the proposed method flags anomalies in two distinctive ways. The MCOD detects distance-based divergences at the end of every time series. The polynomials created over 3-h and 24-h periods, when compared using Frechet distance, reveal any anomalies of actual versus expected behaviour of a cluster.

On the other hand, a deep learning for real-time malware detection has been analysed in [6]. Signature-based approaches form the current industry standard for malware detection, despite obvious shortcomings with detecting obfuscated malware, zero-day exploits and simply the astounding daily number of new malware releases. To tackle these kinds of problems, anomaly detectors based on machine learning models are implemented. Methods like K-nearest neighbour, support vector machines, or decision tree algorithms struggle with high false positive rates. Without sufficient context, malware classification is hard to perform accurately. On the other hand, Deep Learning (DL) algorithms are adequate for making superior decisions, but at a cost. DL needs significantly more time to retrain the detection model, which constitutes a major drawback when new malware strains have to be added frequently. The proposed procedure tries to strike a balance between the accuracy of deep learning and the swiftness of classical machine learning methods by employing a multi-stage detection algorithm cooperating with the operating system. The first stage involves classical machine learning detection. In case the ML classifies a potential threat, it is carried over to the 'uncertain stage'. In the second stage, a Deep Learning algorithm decides if the threat is marked as benign or as hostile, and therefore, killed. If new malware is found, the model is retrained with the use of a concept-drift component, which makes sure the model is relevant.

Fig. 1. The general overview of the proposed system architecture.

3   Proposed System Architecture

The general architecture of the proposed solution has been shown in Fig. 1. Conceptually, the data is collected from the network in a NetFlows format. It captures aggregated network properties. Commonly, that kind of data is collected by network elements and later sent to the collectors. The statistics retrieved from NetFlows are often used by network administrators for auditing purposes. Single NetFlow aggregates statistics (e.g. number of bytes sent and received) about packets that have been sent by specific source address to a specific destination address. Usually to capture the long-term malicious behaviour of a specific node, additional analysis of the data is required. In the proposed approach we calculate statistical properties of a group of NetFlows that have been collected for a specific source IP address within a fix-length time spans called time windows. These statistics include:

- number of NetFlows,
- number of source ports,
- number of protocols used,
- number of destination IP addresses,
- number of destination services,
- sum of bytes exchanged between source and destination,
- total number of exchanged packets.

4   Sparse Autoencoder Overview

The Autoencoder is a type of an artificial feedforward neural network that is trained in an unsupervised manner. During training, the target values are set to be equal to the input values and the backpropagation algorithm is utilised. The architecture of the network consists of three layers – one hidden layer, and two visible ones (see Fig. 2).

Typically, for classical Autoencoder, the number of hidden neurons is lower than the number of neuron in visible layers, so that the network is trained to

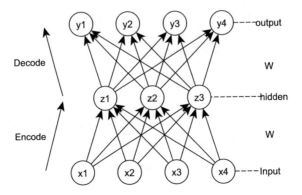

Fig. 2. The structure of the auto encoder.

learn the low-level representation of the underlying data. In case of a Sparse Autoencoder, the number of hidden neurons is significantly higher than in the visible layers. In such case, the network is trained to learn a sparse representation. Typically, bringing the data into a higher dimension reveals interesting facts about the data and allows achieving better classification results. Typically, for the Autoencoders the tied weights concepts is used. It means that the neural network weights for encoding and decoding are the same. The responses of hidden units are calculated using formula (1), where σ indicates sigmoid activation function $(\sigma(x) = \frac{1}{1+exp(-x)})$, N number of visible units, M number of hidden units, c hidden biases, b output layer biases, and w the weights of the artificial neural network.

$$z_j = \sigma \left(\sum_{i=1}^{N} w_{ij} x_i + c_j \right) \tag{1}$$

The input data is reconstructed using (2) formula. The parameters to be trained are weights w and biases vectors b and c.

$$y_i = \sigma \left(\sum_{i=1}^{M} w_{ji} z_j + b_i \right) \tag{2}$$

In order to train the Autoencoder the negative log likelihood loss function is optimized with respect to w, b and c. The loss function has following form:

$$E(x_i, y_i) = - \sum_{i=1}^{N} [x_i ln(y_i) + (1 - x_i) ln(1 - y_i)] \tag{3}$$

Calculating $\frac{\partial E}{\partial w_{ij}}$ for such defined loss function, it is easy to show that weight in k-th iteration can be updated using following formula:

$$w_{ij}^{(k+1)} = w_{ij}^{(k)} + \gamma \left\{ \left(\sum_{i=1}^{N} w_{ij}^{(k)} (x_i - y_i) \right) z_j (1 - z_j) x_i + (x_i - y_i) z_j \right\} \tag{4}$$

where γ indicates learning rate. Similarly, we calculate updates of biases, using following formulas:

$$c_j^{(k+1)} = c_j^{(k)} + \gamma \left\{ \left(\sum_{i=1}^{N} w_{ij}^{(k)}(x_i - y_i) \right) z_j(1 - z_j) \right\} \qquad (5)$$

$$b_i^{(k+1)} = b_i^{(k)} + \gamma(x_i - y_i) \qquad (6)$$

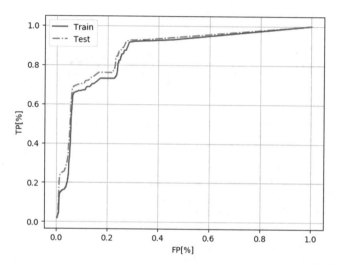

Fig. 3. Effectiveness of detected anomalies (TP) versus number of false positive alerts (FP).

5   Experiments

In order to evaluate the proposed solutions, we have used datasets provided by Malware Capture Facility Project [1]. The datasets are fully labelled and each corresponds to different scenarios of malware infections and/or botnet activities. The dataset has been divided into training and evaluation parts. In practice, the detection algorithm is trained on a subset of scenarios and tested on the others. It is a more real-life strategy than a commonly used approach where part of a specific scenario is used for training and the remaining part for testing. In order to tune the parameters of the model we additionally split the training dataset into 5 parts and performed 5-fold cross-validation.

It must be noted that we perform system training only on the normal traffic. The autoencoder is explicitly never exposed to the traffic samples which include malicious behaviour. This simplifies the learning process since the administrator does not have to label the data. The experiments showed that the proposed

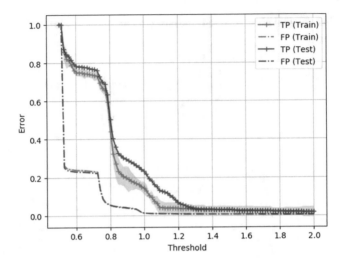

Fig. 4. Effectiveness metrics (for training and validation datasets) versus detection threshold.

solution can achieve 80% of detection effectiveness while having 20% of false positives. The results are comparable both for the data we used for training and for unknown traffic we used for validation (testing).

The average performance of the proposed method has been shown in Fig. 3. It can be noticed that both for training and testing datasets the results are comparable. Moreover, the errors (True Positive and False Positives Rates) with respect to the changing threshold have been shown in Fig. 4. The green band on the figure indicates the range of three standard deviations for the training dataset.

6   Conclusions

In this paper, we have demonstrated the proposal of the unsupervised machine learning approach to cybersecurity. The proposed solution has been used to detect the malicious behaviour of hosts in the network. We have analysed a specific demonstration use case, where the problem of botnet detection by means of analysing the data in form of NetFlows is considered. The presented results are promising.

References

1. The Malware Capture Facility Project. https://mcfp.weebly.com/
2. Grill, M., Nikolaev, I., Valeros, V., Rehak, M.: Detecting DGA malware using Net-Flow. In: IFIP/IEEE International Symposium on Integrated Network Management (IM), Ottawa, pp. 1304–1309 (2015). https://doi.org/10.1109/INM.2015.7140486

3. Abt, S., Baier, H.: Towards efficient and privacy-preserving network-based botnet detection using NetFlow data. In: Proceedings of the Ninth International Network Conference (INC 2012) (2012)
4. Tran, Q.A., Jiang, F., Hu, J.: A real-time NetFlow-based intrusion detection system with improved BBNN and high-frequency field programmable gate arrays. In: IEEE 11th International Conference on Trust, Security and Privacy in Computing and Communications, Liverpool, pp. 201–208 (2012). https://doi.org/10.1109/TrustCom.2012.51
5. Flanagan, K., Fallon, E., Awad, A., Connolly, P.: Self-configuring NetFlow anomaly detection using cluster density analysis. In: 19th International Conference on Advanced Communication Technology (ICACT), Bongpyeong, pp. 421–427. https://doi.org/10.23919/ICACT.2017.7890124
6. Yuan, X.: PhD forum: deep learning-based real-time malware detection with multistage analysis. In: IEEE International Conference on Smart Computing (SMARTCOMP), Hong Kong, pp. 1–2 (2017). https://doi.org/10.1109/SMARTCOMP.2017.7946997

Network Self-healing

Emilia Voicu and Mihai Carabas[(⊠)]

University Politehnica of Bucharest, Bucharest, Romania
emilia.voicu@cti.pub.ro, mihai.carabas@cs.pub.ro

Abstract. A major issue for growing networks is maintenance: as the amount of work increases, manual management becomes impossible. To ensure peak performance and avoid the propagation of errors that could potentially bring down the entire network, the project aims to design a centralized network monitoring tool which would detect anomalies such as increased loss or latency, isolate source of error and take appropriate action to ensure the network maintains its integrity and availability, as well as reduce the mean time to repair. The project is designed to work for the network inside a data center and it looked into different strategies that would fit the current network topology.

1   Introduction

Network management is a vital aspect of maintaining the peak performance of different services. Although in recent years the main trend for infrastructure networks seems to be the moving to the cloud [1], the cost of maintaining your private cloud is significantly lower, on-premise being from 5 to 7 times cheaper [2], leading to many of the companies with a need for large scale networks to build infrastructure or use a combination of the both private and public cloud. One of the recent trends is focusing on reducing the time of detection and repair through automation and orchestration [3].

Objectives. The main objective of the project is to automate the first response to network errors by triggering action as soon as they occur [4]. A data center can hold hundreds of nodes and maintain several services - Fig. 1 shows a small-scale model of the target network - each component is either a router or a layer 4 switch. The healing procedure prioritizes accuracy should not downgrade the network and has a reduced scope in order to minimize the damage caused by false positives. Reducing the detection time and applying small patches to the most common issues aims to increase the scalability and availability as well as decrease the costs induced by failures.

2   The Solution

The solution consists of three main modules that tackle a different problem: monitoring, error detection and healing. The first step is identifying the symptoms

© Springer Nature Switzerland AG 2019
M. Choraś and R. S. Choraś (Eds.): IP&C 2018, AISC 892, pp. 200–207, 2019.
https://doi.org/10.1007/978-3-030-03658-4_24

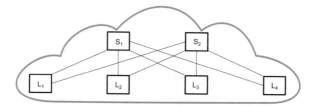

Fig. 1. Small-scale model of data center network

- the factors that indicate there is something malfunctioning inside the network: the real symptoms of a fault are increased latency and increased package loss ratio [5]. The next step is analyzing the collected data, running it against the current thresholds in order to extract a cause and, eventually, undergo repairing routines - most often these routines only trigger an alarm and a human hypervisor needs to manually look into the issue. Figure 2 depicts the architectural overview of the tool as well as all the dependencies between the different components.

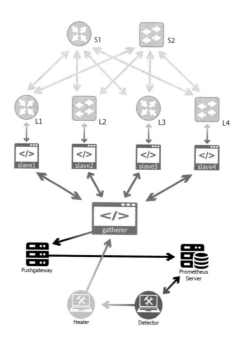

Fig. 2. Architectural overview

2.1   Data Gathering

The data gathering module is the first module to start up in the tool and it should run a few iterations before the other tools start working in order to provide them with enough data to produce relevant output. The overview presents a server-client model in which the `gatherer` is the server and the `slaves` are the clients. When a `slave` starts up, it connects to the master and sends a hosts request, asking the master to send it a list of other Slaves to monitor and start sending `mytraceroute` requests with the `report` and `c 1` to each of those other slaves in its data center, encapsulates the results into JSON format and sends it back to the Gatherer.

Different Monitoring Tools. The data gathering algorithm is not restricted by the type of tool used - mtr, fbtracert or plain traceroute. However, each of these has a different performance and will complete faster or later, which forces the Slave to spawn more child processes that run the command with one second delay in order to provide data for the Gatherer each second. The size of the network is also taken into account, as well as the hardware capabilities of the Slave, when designing the gathering algorithm. The following measurements were observed in the small-scale network described in the previous section (Table 1).

Table 1. Correlation between tool, duration and number of processes

Tool	Duration	Number of processes
mtr	6 s	6
fbtracert	0.8 s	1
traceroute	30 s	12[a]

[a]12 is a hardcoded value in the slave that forbids it from overloading and crashing. It induces the risk of slowing down the overall performance of the tool but the gain comes from increased accuracy - the number of active slaves is direct proportional to the accuracy of the tool

2.2   Error Source Detection

The gatherer formats this data into the appropriate format to be published into the `Pushgateway` and then have it scraped by `Prometheus`. The Detector module queries the Prometheus database for loss and latency so as to detect any issues. Two methods have been investigated for detecting an issue: look at each device and determine its sanity and report it if problematic or look into links and extrapolate to devices where necessary.

Device-Oriented Algorithm. For this method, each slave will look at each device and count the faulty links. If the number of links is greater than a given thresholds - i.e. 75% of links, then the slave reports the device as problematic. This algorithm prioritizes the discovery of problems with the risk of increased false positives. One scenario where this algorithm is desirable is when the healing procedure has mechanisms that can reverse any damage inflicted by the healing routines when applied for false positives. However, a heavier load is put on the healer since the number of reported devices increases greatly and the constant return of the network from an induced degraded state to a previous healthy one degrades the tool's performance. One major downside of the algorithm is that it does not take into account propagated errors - if a link fails, the rest of the links on that path will appear to have errors even if they are healthy (Fig. 3).

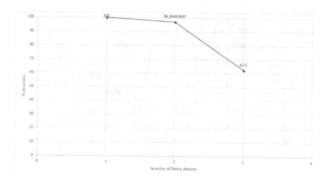

Fig. 3. Algorithm's accuracy when increasing the number of faulty devices

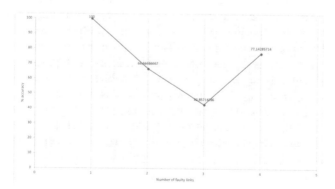

Fig. 4. Algorithm's accuracy when increasing the number of faulty links

Performance. The measurements were made on the small-scale topology presented in the previous chapters. The poor accuracy comes from the large number of false positives reported and not from unreported incidents or links.

Link-Oriented Algorithm. In the first step, each slave will look at each path towards the other slaves and mark the first link that has errors with the colour orange and all the other links in the path that present errors, after that one, are marked with the colour yellow. At the end of this step, there will be a list of orange links reported by each slave. If a link is reported by less than 50% of the slaves, the yellow list is checked and the ones reported by multiple slaves are moved to orange. Afterwards, the link is moved to red, as presenting a real issue. When the orange list has been emptied, a routine is applied to the red list to extract problematic devices as follows: find the devices with the most occurrences in the links, if more than 75% of their links are in the red list, then report the device as faulty. In case of faulty devices, this algorithm reports no false positives. However, there are instances where it will not report all the faulty devices at once - if two devices interfere with the other's measurements then only one will be reported. The reporting of multiple devices happens when the two devices are not on the same path, otherwise the detection process takes multiple stages. A couple of edge cases arise: when 75% percent of a device's links are indeed malfunctioning but the device is healthy - highly unlikely in a data center where the sanity of the network is often 100%, and when the links between the Slaves and the network devices fail. The algorithm checks for ill-reporting slaves - if something in the orange list is only reported by a single slave and it does not appear in the yellow list, then the slave has a problem (Fig. 4).

2.3 Healing

Once the problem has been identified, the relevant data is sent to the healer. The healer can determine what the device's or link's issue is. It has a set of actions that it undergoes for specific problems. Once the problem has been isolated - too much CPU usage, a flapping interface, etc. - a set of commands is issued that are meant to either fix the problem or increase the path's cost so as to remove it without risking to bring down the network in the case of multiple devices behaving that way - in the worst case scenario, if all the devices are turned off, the network shuts down whereas just increasing the path cost keeps the network up, even if performing poorly. The commands are then sent to the Gatherer, along with the address of the problematic device which will save them in an internal buffer and send them to the slave instead of the next `ACK`. After receiving the command, the slave connects to the device and runs it. If the problem cannot be identified, instead of sending the healing routine, an alarm is triggered and human intervention is due.

3   Results

The self-healing tool is designed to work in a small-scale data center network where an issue is often treated as soon as it appears. The measurements were obtain by creating a virtual environment of network components such as Nexus switches[1] and simulating the network. Those measurements are theoretical and do not represent the performance the tool would have in a real network since it did not consider the workload of the network.

3.1   Speed

The processing time - data processing and detection algorithms, makes up for a very small part of the total time, most of it comes from the detector which is set to query the Prometheus data base every 10 s. The number of seconds ensures it waits for enough data to be published so as to have a consistent amount of queries on which to work - the gatherer receives data from a slave every second. Depending on the detection tool, the frequency with which the gatherer publishes data can vary. The desired frequency is of one second, regardless of the method used, that is why the Slave requires parallelization. The total time that passes from the moment an issue occurs and the moment it is resolved is given by the equation:

$$t_{total} = t_{publish} + t_{query} + t_{heal} \tag{1}$$

$t_{publish}$ = time from the moment the issue occurs until it is published

$$t_{publish} = t_{traceroute} \tag{2}$$

t_{query} = time from publishing to the detector finding it

$$t_{query} = 10 - t_{last\_query} \tag{3}$$

t_{heal} = time from the moment the issue is reported until it is healed

$$t_{heal} = 6 - max(message\_number) + t_{healing\_routine} \tag{4}$$

3.2   Correctness

A source of errors for the algorithm is the detection of a problematic device. Since there is no direct access to network components for detection, a malfunctioning device is reported by looking at all its links and if the number of problematic links associated with it is greater than 75% of its total number of links, the device is reported as having a problem instead of reporting a list of all its links. Problems arise in networks where the probability of multiple links failing at the same time is greater than the probability of devices failing. The correctness of the device-oriented algorithm drops significantly when multiple issues appear on the same level. If both S_1 and S_2 are faulty by having all the links associated with them faulty at the same time, the detector outputs:

---

[1] https://www.cisco.com/c/en/us/products/switches/data-center-switches/index. html#~stickynav=2.

```
Detected faulty devices: {'S1', 'S2', 'L1', 'L2', 'L3'}
Detected faulty links:
{'slave1': {('L3', 'slave3'), ('L2', 'slave2')},
 'slave2': {('L3', 'slave3'), ('L1', 'slave1')},
 'slave3': {('L1', 'slave1'), ('L2', 'slave2')}}
```

Using the device-oriented algorithm for the same scenario, the `detector` has 100% accuracy and no risk of false-positives:

```
Red devices: {'S1', 'S2'}
Orange devices: {}
Yellow devices: {}
Links: {}
```

In extreme cases, where the links turn out to be faulty instead of the devices, the correctness suffers a significant drop. However, for the current topology, that would imply that 66% of all the links in the network should suffer a problem at the same time and the chances of that happening are too slim to be an issue.

3.3 Limitations

The current status of the project has no integrity-check as it relies on the following iterations of the tool to detect whether the error is still present or if another error is present - introduced by the tool or not. A feature that could be added is a mechanism for snapshots - saving the current state of the network and restoring it on demand as well as a mechanism for performance checks which can alert on inappropriate healing routines that reduced the network's performance. Because most data centers already have monitoring tools, data from those tools can be extracted and taken into account in order to increase the accuracy - i.e. the machine-centric data can be interrogated when it comes to detecting whether links or devices are at fault. Another limitation is the fact that it does not tackle issues that occur between data centers.

4 Conclusions

The project investigated the behavior of a data center network and looked into several factors that cause problems and methods of intervention that would not affect the network's overall behavior. The most relevant ones were latency and loss ratio - they had the most significant impact on applications relying on the network that is why the chosen tools to employ for monitoring were `traceroute` and `ping` which offer the desired measurements. The `gatherer` parses all this data and publishes it to Prometheus from where the detection tool takes it and runs detection algorithms on it. Once the fault has been isolated, the healing mechanism is triggered and it launches the healing routines appropriate for the type of fault detected.

The observed results suggest that the topology of the network is a vital part of the implementation and it greatly affects the tool's behavior. One important aspect is the alarming on false positives which can result in network downgrading. However, the link-oriented algorithm, with a more complex approach and a multi-stage detection method minimizes the cases for those occurrences. The speed of reaction is tied to the constraints imposed by the network and it can be altered depending on the desired results.

Acknowledgement. This work was supported by SIMULATE project (no. 15/17.10.2017): Simulations of Ultra-High Intensity Laser Pulse Interaction with Solid Targets.

References

1. McLellan, C.: Cloud v. data center: key trends for IT decision-makers. ZDNet (2018). https://www.zdnet.com/article/cloud-v-data-center-key-trends-for-it-decision-makers/. Accessed 18 June 2018
2. Mytton, D.: Saving $500k per month buying your own hardware. Cloud vs colocation. Server Density Blog (2018). https://blog.serverdensity.com/cloud-vs-colocation/. Accessed 18 June 2018
3. Savage, M.: 4 emerging networking trends. Network computing (2017). https://www.networkcomputing.com/data-centers/4-emerging-networking-trends/728900591. Accessed 19 June 2018
4. Paessler, D.: Interview: the benefits of network monitoring (Part 1). Blog-paesslercom (2018). https://blog.paessler.com/interview-the-benefits-of-network-monitoring-part-1. Accessed 20 June 2018
5. Rogier, B.: Measuring network performance: latency, throughput and packet loss. Accedian (Performance Vision) (2018). https://www.performancevision.com/blog/measuring-network-performance-links-between-latency-throughput-packet-loss/. Accessed 19 June 2018

The Rough Set Analysis for Malicious Web Campaigns Identification

Michał Kruczkowski and Mirosław Miciak[✉]

Faculty of Telecommunications, Computer Science and Electrical Engineering,
UTP University of Science and Technology, Bydgoszcz, Poland
{michal.kruczkowski,miroslaw.miciak}@utp.edu.pl

Abstract. The malware (malicious software) identification has to be supported by a comprehensive and extensive analysis of data from the Internet, and it is a hot topic nowadays. Data mining methods are commonly used techniques for identification of malicious software. Multi-source and multi-layered nature of propagation of malware assumes the utilization of data from heterogeneous data sources, i.e., data taken from various databases collecting samples from multiple layers of the network ISO/OSI reference model. Unfortunately, analysis of such multi-dimensional data sets is complex and often impossible task. In this paper we investigate multi-dimensionality reduction approach based on rough set analysis.

1 Introduction

Recently, numerous attacks have threatened the operation of the Internet [3]. One of the major threats on the Internet is malicious software, often called malware. Malware is a software designed to perform unwanted actions on computer systems such as gather sensitive information, or disrupt and even damage computer systems. In case of finding a server vulnerability (similar to the disease [6,17]) affecting multiple web pages, the attacker can automate the exploitation and launch an infection campaign in order to maximize the number of potential victims. Hence, the campaign determines a group of incidents that have the same objective and employ the same dissemination strategy [4,5,14]. Detecting malicious web campaigns has become a major challenge in Internet today [3,7,8,11]. Therefore, strategies and mechanisms for effective and fast detection of malware are crucial components of most network security systems.

In this paper we propose the use of rough set analysis for dimensionality reduction of analyzed data sets. It can be important in the case of big data sets classification. In this case the reduction of dimensionality can significantly accelerate the classification process. From practical point of view it is often impossible to classify such collections on account of their size, multi-dimensionality and huge computing complexity.

The analysis we investigate based on data sets taken from various databases collecting samples from various layers of the network ISO/OSI reference

© Springer Nature Switzerland AG 2019
M. Choraś and R. S. Choraś (Eds.): IP&C 2018, AISC 892, pp. 208–215, 2019.
https://doi.org/10.1007/978-3-030-03658-4_25

model [10]. Assuming multi-source and multi-layered nature of propagation of malware. The evaluation of the proposed method for data preparation we performed based on system for malware campaigns identification [9]. We prove that dimensionality reduction based on rough set analysis has significant influence on a classification quality of malware samples.

The authors claims that the use of dimensionality reduction should help to obtain better classification results with reference to single classification model.

2   The Rough Set Analysis

The proposed method involves choosing parameters with the use of rough set theory. This technique provides methods for determination of the most important attributes of a computer system without losing classification ability. The rough set theory was developed by Zdzislaw Pawlak [1,2,12,13] and is based on the concept of upper and lower approximation of the set, approximating space and models of sets. Measurement data is presented in the form of a table, whose columns correspond to the attributes and rows corresponds to particular objects or states. Such a record is called an information system [16]. Based on rough set theory is a reduct, which is the main part of the information system. As a result of this, it is possible to distinguish all the distinguishable objects from the original set of attributes, hence, reducts are minimal subsets of attributes maintaining characteristics of the whole set of attributes. Literature studies provides many algorithms for determination of sets of information system reducts. For the needs of our solution, the authors have taken into consideration the methods of: Johnson, global, local [15].

The input data set is defined as a finite non-empty set U, also called in the theory of rough sets, a universe with the following properties:

1. elements of the U set are called objects,
2. any family of subsets of the U set, we call knowledge about U,
3. the division of U, i.e. a family of disjoint and non-empty subsets U, such that their sum equals the entire set U, is called the classification of the set U,
4. a pair of $K = (U, C)$, where $C = C1, ..., C_n$ is any family of divisions of the set U, we call the knowledge base about U.

Data are often called samples, records or lines and can be interpreted as states, cases, observations, etc. The information system consists of $B = (U, A)$, where U is the universe, $A = a_1, ..., a_n$ is a finite, non-empty set of attributes (parameters) describing objects, Each attribute $a \in A$ can be interpreted as the assignment $a : U \rightarrow V_a$, where $V \neq \emptyset$ is a set of values of the attribute a. If $A = (U, A)$ will be an information system, then $B = (U', A')$ is the subsystem A, $B \subseteq A$ induces $B' \subseteq A'$ and $A' = A$.

2.1   Multi-dimensionality Reduction

Let $A = (U, A)$ be an information system and $B \subseteq A$. The set of all reducs of the B attribute set in the information system A in the information system A will be marked $RED_A(B)$. In other words, reduct is a subset of attributes, which from the point of view of object distinguishability carries the same information as the whole set of attributes and is also minimal due to this property. The algorithm of determining the set of reducts is as follows: The presented method is based on calculating the matrix of distinguishability M, whose size is square due to the number of objects. Existing approximation methods allow to determine the necessary information about the discrimination matrix directly from the data and do not require the calculation of the entire matrix. Significance of the attributes is described on the basis of the determined set of reducts, where the attributes are ordered according to the number of occurrence in the set.

Algorithm 1. Multi-dimensionality reduction based on rough set

---
1: Input: Information system $A = (U, A)$
2: Output: Set $RED_A(A)$ of all reduction A
3: *Determine the discrimination matrix* $M(A) = (C_{ij})$
4: Reduce M using the extraction laws: d number of non-empty fields of the M
5: Build the family of sets $R_0, R_1, \ldots, R_d$ as follows:
6: begin
7:        $R0 = 0$
8:        for $i = 1$ to d
9:              $R_i = S_i \cup T_i$, where $S_i = R \in R_{i-1} : R \cap C_i \neq 0$ and $T_i = (R \cup \{a\})$
10: end
11: Remove any unnecessary elements for each element of the R_d family
12: For each R_d family element, remove the repeating elements
13: return $RED_A(A) = R_d$

---

3 Data Preparation Taken from Heterogeneous Data Sources

The architecture of data preparation is presented in Fig. 1. It consists of databases collecting malicious data taken from heterogeneous data sources, data pre-processig module and feature extraction module. The analyses determine a structure of a data aggregated in the central database. It can be the data provider to the classification system. We used this data preparation for the malicious web campaigns identification system described in the paper [9]. In this paper malicious software (malware) samples are used. These samples are represented by a vector of attributes (Eq. 1).

$$MS_i = [attribute_{1i}, attribute_{2i}, \ldots, attribute_{ni}, class_i] \tag{1}$$

where MS_i is i-th malware sample and $attribute_{ni}$ represents a value of n-th attribute of i-th malware sample and $class_i$ is the class label of the i-th sample. We considered two class labels: $\{-1\}$ for samples related with malicious campaigns and $\{+1\}$ for samples unrelated with any malicious campaigns. It should be noted that all data samples represent malicious activity and our goal is to assess their participation in massive web campaigns. Unfortunately, access to databases collecting malicious data divided into campaigns (if they exist) is restricted. Therefore, the only sensible solution is to define some patterns related with campaigns based on analysis of existing databases collecting malicious data, and generate a set of samples containing these patterns.

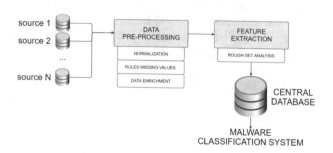

Fig. 1. The architecture of data preparation based on the rough set analysis.

Before the rough set analysis data sets are prepared. It means all attributes are normalized, i.e. each data attribute is represented as a decimal fraction taking the values from $(0, 1)$ interval. Moreover, the important atomic values are extracted from the data.

$$scheme : //domain : port/path?query\_string\#fragment\_id \qquad (2)$$

The URL address (Eq. 2) consists of the following components, i.e., scheme, domain name or IP address, port, path, query string and a fragment identifier (some schemes also allow a username and password in front of the domain). The query string consists of pairs of keys and values (also called attributes). By malicious URL we denote such URL that is created with malicious purposes. It can serve malicious goals, for example download any type of malware, which can be contained in phishing or spam messages, etc. It is obvious that the ultimate goal of obfuscation is to make each URL unique. Due to the fact that malicious URLs are generated by tools which employ the same obfuscation mechanisms for a given campaign we assume that usually intruders keep some parts of the URL static, while other parts are changed systematically and in an automated fashion.

After preliminary analysis of the available data the following attributes have been often selected by rough set: date, time, address (IP, ASN), length of address, domain name, number of subdomains, path name, number of queries, country code, confidence of code.

4   Case Study

4.1   The n6 Platform

The rough set analysis was used to prepare heterogeneous data from a real malware databases from the n6 platform to identify malicious campaigns. The n6 platform [10] developed at NASK (Research and Academic Computer Network) is used to monitor computer networks, store and analyse data about incidents, threats, etc. The n6 database collects data taken from various sources, including security organizations, software providers, independent experts, etc., and monitoring systems serviced by CERT Polska. The data sets contain URLs of malicious websites, addresses of infected machines, open DNS resolvers, etc. Most of the data is updated daily. Information about malicious sources is provided by the platform as URL's, domain, IP addresses, names of malware, etc. We have performed a preliminary analysis of 7 234 560 data samples collected in year 2018 and stored in the n6 database (Fig. 2).

Fig. 2. The n6 platform architecture.

4.2   Case Study Results

Table 1 presents the structure of data sets, ie. number of samples, number of data sources and the contribution of class labels. It should be noted that analyzed data sets consist of imbalanced data labels. In our case study we received more samples unrelated with any campaigns. Such a situation makes it much more difficult to generalize the classification problem. Nevertheless, presented implementation is resistant to unbalanced data sets.

Table 1. The structure of data taken from heterogeneous sources.

Number of samples (URLs)	7 234 560
Number of data sources	32
Percent of unique samples	72 %
Percent of malware samples related with campaigns $\{-1\}$	38 %
Percent of malware samples unrelated with any campaigns $\{+1\}$	62 %

Next, the data sets are prepared for further classification based on rough set analysis according to the Algorithm 1.

The result of rough set analysis is training data set used for data classification. We evaluated our case study based on the SVM classifier presented in the paper [9].

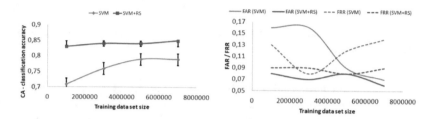

Fig. 3. The classification accuracy and FAR/FRR for variable sizes of input data sets based on cross validation with 5 folds.

The results of the analysis were presented on the Fig. 3 for two configurations (1) SVM system with rough set analysis (2) SVM system without rough set analysis. Classification accuracy (CA) as a ratio of number of correctly identified samples to the size of the analyzed data set with reference to various size of training data set was considered. Classification accuracy can be expressed:

$$CA = \frac{TP + TN}{TP + TN + FP + FN} \tag{3}$$

where TP denotes the number of true positive predictions, TN the number of true negative predictions, FP the number of false positive predictions and FN the number of false negative predictions.

Moreover the False Acceptance Rate (FAR) is calculated as a fraction of false positive predictions FP and false positive and true negative predictions $FP + TN$ as follow:

$$FAR = FPR = \frac{FP}{FP + TN} \tag{4}$$

The False Rejection Rate (FRR) is calculated as a fraction false negative FN by false negative and true positive predictions $FN + TP$ as follow:

$$FRR = FNR = \frac{FN}{FN + TP} \tag{5}$$

Simultaneously we compiled the results of time of building classifier models with reference to various size of training data set on the Fig. 4.

The achieved classification accuracy ranged from about 71% to 79% for classifier without rough set analysis and from about 83% to 85% for classifier with rough set analysis. Time of building classifier models ranged from about 19 min to 91 min for classifier without rough set analysis and from about 3 min to 11 min for classifier with rough set analysis.

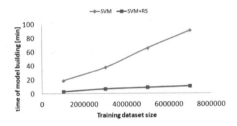

Fig. 4. The time of classification for variable sizes of input data sets based on cross validation with 5 folds.

In general, presented results confirm that the accuracy of classification and time needed for building a predictive model clearly depends on the size and dimensionality of training data sets. Moreover, it is very important to choose the adequate classification methods.

5   Conclusions

Data mining methods especially multi-dimensionality reduction hold a great potential toward detection of malicious software. To identify malicious web campaigns we have adopted a rough set analysis and supervised learning methods. In general, the presented results of experiments confirm that application of rough set analysis has strong influence for classification results, i.e. accuracy, time needed for building classifier model. Authors claimed that multi-dimensionality reduction in long term analysis can reduce the impact of classifier model overfitting. It means that the models have greater ability to generalize the classification problem. This analysis can be successfully used to analyse a huge amount of dynamic, heterogenous, unstructured and imbalanced network data samples. As a final conclusion we can say that it can be expected that our approach can be successfully implemented in intrusion detection systems as a malicious campaign sensor.

References

1. Bazan, J., Szczuka, M., Wojna, A., Wojnarski, A.: On evolution of rough set exploration system. In: Lecture Notes in Articial Intelligence, pp. 592–601 (2004)
2. Bazan, J., Szczuka, M., Wróblewski, J.: A new version of rough set exploration system. In: Lecture Notes in Articial Intelligence, pp. 397–404 (2002)
3. Gandotra, E.: Malware analysis and classification: a survey. J. Inf. Secur. **5**, 56–64 (2014)
4. Gao, H., Hu, J., Wilson, C., Li, Z., Chen, Y., Zhao, B.: Detecting and characterizing social spam campaigns. In: Proceedings of the 10th ACM SIGCOMM Conference on Internet Measurement, pp. 35–47 (2010)
5. Hajmasan, G., Mondoc, A., Cret, C.: Dynamic behaviour evaluation for malware detection. In: 2017 5th International Symposium on Digital Forensic and Security (ISDFS) (2017)

6. Kasprzyk, R., Najgebauer, A., Pierzchała, D.: Modelling and simulation of an infection disease in social networks. In: Computational Collective Intelligence. Technologies and Applications, vol. 1, pp. 388–398. Springer (2011)
7. Kruczkowski, M., Niewiadomska-Szynkiewicz, E.: Support vector machine for malware analysis and classification. In: Proceedings of the IEEE/WIC/ACM International Conference on Web Intelligence, pp. 1–6 (2014)
8. Kruczkowski, M., Niewiadomska-Szynkiewicz, E., Kozakiewicz, A.: Cross-layer
   analysis of malware datasets for malicious campaign identification. In: Proceedings of the International Conference on Military Communications and Information
   Systems (2015)
9. Kruczkowski, M., Niewiadomska-Szynkiewicz, E., Kozakiewicz A.: FP-tree and
   SVM for malicious web campaign detection. In: Proceedings of the 7th Asian
   Conference on ACIIDS 2015. Part II. Lecture Notes in Computer Science, Bali,
   Indonesia, 23–25 March 2015, vol. 9012, pp. 193–201 (2015)
10. NASK: n6 platform (2014). http://www.cert.pl/news/tag/n6
11. de Oliveira, I.L., Ricardo, A., Gregio, A., Cansian, A.: A malware detection system
    inspired on the human immune system. In: LNCS. vol. 7336, pp. 286–301. Springer
    (2012)
12. Pawlak, Z.: Rough sets. Int. J. Parallel Program. 341–356 (1982)
13. Pawlak, Z.: Information systems - theoretical foundations (1983)
14. Rahman, M., Rahman, M., Carbunar, B., Chau, D.H.: Search rank fraud and
    malware detection in google play. IEEE Trans. Knowl. Data Eng. (2017)
15. Salamo, M., Golobardes, E.: Global, local and mixed rough sets case base main-
    tenance techniques. In: Recent Advances in Artificial Intelligence Research and
    Development, pp. 217–234 (2004)
16. Walczak, B., Massart, D.: Rough sets theory. Gemoemet. Intell. Lab. Syst. **47**(1),
    1–16 (1999)
17. Yuan, X.: Deep learning-based real-time malware detection with multi-stage analy-
    sis. In: 2017 IEEE International Conference on Smart Computing (SMARTCOMP)
    (2017)

Assessing Usefulness of Blacklists Without the Ground Truth

Egon Kidmose[1,2(✉)], Kristian Gausel[2], Søren Brandbyge[2],
and Jens Myrup Pedersen[1]

[1] Department of Electronic Systems, Aalborg University, Fredrik Bajers Vej 7,
9220 Aalborg Øst, Denmark
{egk,jens}@es.aau.dk
[2] LEGO System A/S, Aastvej, 7190 Billund, Denmark
{egon.kidmose,kristian.gausel,soeren.brandbyge}@LEGO.com

Abstract. Domain name blacklists are used to detect malicious activity
on the Internet. Unfortunately, no set of blacklists is known to encompass
all malicious domains, reflecting an ongoing struggle for defenders to keep
up with attackers, who are often motivated by either criminal financial
gain or strategic goals. The result is that practitioners struggle to assess
the value of using blacklists, and researchers introduce errors when using
blacklists as ground truth. We define the ground truth for blacklists to
be the set of all currently malicious domains and explore the problem
of assessing the accuracy and coverage. Where existing work depends on
an oracle or some ground truth, this work describes how blacklists can
be analysed without this dependency. Another common approach is to
implicitly sample blacklists, where our analysis covers all entries found
in the blacklists. To evaluate the proposed method 31 blacklists have
been collected every hour for 56 days, containing a total of 1,006,266
unique blacklisted domain names. The results show that blacklists are
very different when considering changes over time. We conclude that it
is important to consider the aspect of time when assessing the usefulness
of a blacklist.

Keywords: Domain names · Blacklists · Domain Name System

1 Introduction

Domain names are crucial for a lot of Internet activity, including malicious activities of cyber criminals. Examples of such use include criminals sending SPAM and phishing emails, where domains are used for linking to malicious resources, or it can be observed when malware successfully infects a victim machine and needs to establish a Command and Control (CnC) channel with the attacker. This is often achieved through the Domain Name System (DNS).

If malicious domains are known, queries to resolve them through DNS provides for simple detection. Blocking the responses can prevent or inhibit malicious activity. Consequently, practitioners employ domain name blacklists to

© Springer Nature Switzerland AG 2019
M. Choraś and R. S. Choraś (Eds.): IP&C 2018, AISC 892, pp. 216–223, 2019.
https://doi.org/10.1007/978-3-030-03658-4_26

defend systems. Blacklists also finds use in research on detecting malicious domain names, and malicious activity in general. While the research area is very much active and goes beyond using blacklists for detection, blacklists are still used as a ground truth and for validation [10].

Unfortunately, the attackers hold the initiative and choose which domain they register and which they abuse, making it hard to determine if a domain is malicious or not, and even harder to identify all malicious domains. This leads to a problem of lacking coverage, i.e. the fact that the known malicious domains only cover a subset of the ground truth of all malicious domains [9]. In addition to this, criminals have adopted their practices to exploit any time lag until a domain is blacklisted, making timeliness a challenge with further negative impact on accuracy [5,11]. In summary, the accuracy of blacklists is expected to be impeded by insufficient timeliness and coverage, while the ground truth is not available for measuring the accuracy directly.

This paper contributes by presenting a method for assessing the coverage and timeliness of blacklists. An important, novel point is that this is done without assuming any ground truth. Instead, blacklists are analysed according to a set of proposed metrics, that covers all entries of the lists. Ultimately, this paper improves our understanding of benefits and shortcomings of blacklists, considering complete blacklists and without relying on a ground truth.

2   Related Work

Sinha et al. have studied four IP blacklists and their spam detection accuracy. The SpamAssassin spam detector is validated against manually curated spam and non-spam emails and used as an oracle. Results include non-negligible false negative rates (never less than 35%) and false positive rates (as high as 11%). These errors can either be due to lacking blacklist accuracy or due to errors by the oracle. Our work differs by having domain blacklists as subjects and by eliminating the risk of errors by a non-perfect oracle.

A study of blacklists accuracy and timeliness has been conducted by Sheng et al., addressing phishing URLs [9]. 191 newly received phishing URLs are submitted to eight different web browsers/browser plugins, mainly relying on URL blacklisting. Results show that when the phishes are first received, coverage is typically 10% and after two hours most blacklists still cover less than 50%. We expand these insights by observing the full blacklists, doing so for 56 days, thereby processing 1,006,266 unique malicious domains.

There is ample of work on detecting malicious domains that relies on blacklists, either as a data source or as a ground truth, of which we now describe a select few. Antonakakis et al. presents Notos, a system for detecting malicious domains, based on DNS traffic. It relies on three blacklists for ground truth [1]. The nDews system has a similar scope to Notos but identify suspicious domains [6], relying on three blacklists to qualify suspicion in a second phase. FluxBuster is a proposal for detecting malicious domains that exhibit fluxing - a technique applied by criminals to improve resilience of their CnC infrastructure [8]. It is found that malicious domains can be detected days or weeks before appearing in any of 13 blacklists, highlighting the problem of timeliness. BotGAD is a system for detecting CnC domains based on DNS traffic combined with three blacklists [2]. The state of the art for detecting malicious domains is perhaps best captured by PREDATOR, which aims to detect SPAM domains before they are registered, and therefore before they can be abused in any way [4]. The authors behind PREDATOR state that there is a lack of ground truth on malicious domains but find that blacklists are the best available solution for evaluation. Two blacklists make up the ground truth used for evaluation.

Studies on domain abuse also rely on blacklists. Felegyhazi et al. use a blacklist as seed for inferring new malicious domains [3]. Again, the issues of timeliness and lacking coverage are identified. Vissers et al. analyse the modus operandi of cyber criminals, in particular how they register domain names, using blacklists to validate maliciousness [11]. They too confirm challenges regarding timeliness and find that 18.23% of the malicious domains they identify are never captured by the three blacklists in use.

Overall, we find that blacklists are used both for operational purposes and in research, making the problems relating to coverage and accuracy highly relevant. The existing work on assessing blacklists either rely on an oracle being available to provide the ground truth or rely on a source that samples the ground truth for a subset.

In this work we seek to improve the state of the art on two important points:

1. We present a method to analyse entire blacklists, rather than assuming a limited view based on sampling. This provides insights that would otherwise be impossible and eliminates any bias due to sampling.
2. The method does not rely on a ground truth or an oracle to be available. Consequently, errors made by non-perfect ground truth/oracles are eliminated, and so is any need for manually vetting ground truth/oracles.

3   Methods

This section holds a description of the data collection procedure, followed by a description of the proposed metrics.

3.1   Data Collection

A common practice in prior work is to analyse blacklists by sampling them according to some feed of positive samples (e.g. domains found in received spam). The motivation for this choice is unknown but could be due to a preference for Block Lists served through DNS (DNSBL) or other *query-only blacklists*, which has some operational benefits. Drawbacks of using query-only blacklists include that timeliness only can be explored for the time after a domain is produced by the feed of positive samples, and that any bias from the feed will propagate to how the blacklists are sampled. In order to eliminate the sampled view of blacklists, and instead gain full insights into all the blacklisted domains, we focus our study on *retrievable blacklists*. This allows us to analyse all blacklisted domains and not only those hit by sampling. It will also allow us to establish when a domain is blacklisted.

Some blacklist feeds are offered as paid services, likely to cover for the effort of curating blacklists and justified by the value of accurate detection. At the same time, there are communities that have an interest in blacklists, and possibly for that reason there are also free and public community curated blacklists. This work is based on freely available blacklists from both companies and communities, subject to agreeable usage terms.

With our focus being on domain names, we consider domain and URL blacklists, where domains can be extracted from the URLs. Blacklists can be targeted at certain types of maliciousness, but we see no reason to exclude certain types.

Based on the above criteria, a manual search of the Internet has been conducted, with an outset in the blacklists found in related work. The result is a set of 31 retrievable blacklists, targeting either malicious domain names or URLs. The blacklists are retrieved hourly with a system based on MineMeld [7]. Table 1 holds the URLs for each blacklist.

Table 1. Overview of blacklists included in the analysis. BID: Blacklist Identifier. Abbreviations used in in URLs: `https://ransomwaretracker.abuse.ch/downloads` (`<RWT>`), `https://zeustracker.abuse.ch/blocklist.php?download=baddomains` (`<ZEUS>`), `https://hosts-file.net` (`<HF>`), `http://malware-domains.com/files` (`<MD>`) and `http://www.malwaredomainlist.com/mdlcsv.php` (`<MDL>`).

BID	Data feed URL	BID	Data feed URL
1	`<RWT>/CW_C2_DOMBL.txt`	17	`<HF>/grm.txt`
2	`<RWT>/CW_C2_URLBL.txt`	18	`<HF>/hfs.txt`
3	`<RWT>/LY_C2_DOMBL.txt`	19	`<HF>/hjk.txt`
4	`<RWT>/LY_DS_URLBL.txt`	20	`<HF>/mmt.txt`
5	`<RWT>/LY_PS_DOMBL.txt`	21	`<HF>/pha.txt`
6	`<RWT>/TC_C2_DOMBL.txt`	22	`<HF>/psh.txt`
7	`<RWT>/TC_C2_URLBL.txt`	23	`<HF>/pup.txt`
8	`<RWT>/TC_DS_URLBL.txt`	24	`<HF>/wrz.txt`
9	`<RWT>/TC_PS_DOMBL.txt`	25	`http://malc0de.com/bl/ZONES`
10	`<RWT>/TL_C2_DOMBL.txt`	26	`<MDL>`
11	`<RWT>/TL_PS_DOMBL.txt`	27	`<MD>/db.blacklist.zip`
12	`<ZEUS>`	28	`<MD>/immortal_domains.zip`
13	`<MD>/ad_servers.txt`	29	`<MD>/justdomains.zip`
14	`<HF>/emd.txt`	30	`<MD>/malwaredomains.zones.zip`
15	`<HF>/exp.txt`	31	`<MD>/spywaredomains.zones.zip`
16	`<HF>/fsa.txt`		

3.2 Metrics

With a set of 31 blacklists, containing a total of 1,006,266 unique domains, it is evident that some statistics or metrics are required to gain insights. The following is our proposals for metrics to use for such analysis.

As a component to coverage, we consider the **size** as the number of domains found in a blacklist. With malicious domains appearing rapidly, entries must be added to maintain coverage, thus we identify **appearances**. To avoid false positive for once-bad-now-benign domains, domains must be removed so we count **removals**. Finally, removing a malicious domain by mistake is problematic, so we look for **reappearances**, where domains are removed and reappear on the same blacklist, although this also can occur for other reasons.

4 Results

Data collection started on January 25th, 2018, and up to March 22th (56 days) all blacklists found in Table 1 were retrieve every hour. A total of 1,006,266 malicious domains was recorded. During the first day, across all lists, 731,118 domains appeared, while the number was 208,413 for the second day. This makes them the two days with the most appearances, while the 21st day of our collection was the third busiest day, with only 22,267. We assume this to be an effect of a backlog for ingesting already blacklisted domains into our data collection system and define our epoch to start with the third day, March 27th. Figure 1 presents the appearances after the epoch.

A breakdown of size, appearances, removals, and reappearances per blacklist is shown in Table 2. Only 13 reappearances occurred and they all occurred for BID 13 on April 14th. Out of the 31 blacklists, 19 saw no changes.

Domain appearances/removals per list are shown in Figs. 2 and 3.

5 Discussion

This work is, to the best of our knowledge, the most comprehensive study of domain blacklists in terms of number of blacklists and the duration of the observation. It has indeed provided relevant insights into blacklists and their usefulness. In the following we will highlight the most relevant insights.

One important observation is that 19 of 31 blacklists did not change during the 56 days. These 19 blacklists account for only 43,451 of all the 1,006,266 unique domains, meaning the static lists are small in size. Some of the largest static list (BID 15 with 12,306 domains, BID 20 with 5,521 and BID 24 with 3,638) can be summarised as domains where malware and exploits are peddled, or where misleading marketing practices are applied. As such threats are harder to prosecute than direct attacks, it is to be expected that they only change infrequently, and these lists could very well be useful. As for the multiple static malware tracking lists, it seems more questionable if they are useful countermeasures, keeping in mind that criminals have high agility when it comes to using

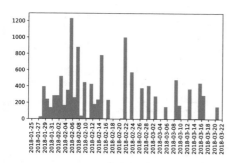

Fig. 1. Number of new domains appearing in a blacklist per day.

Table 2. Size, appearances (App.) and removals (Rem.) by blacklist. Changes (Ch.) is the sum of App., Rem., and Reappearances. Sorted by number of changes. 19 blacklists with no changes are omitted.

BID	Size	App.	Rem.	Ch.
31	4528	4528	4528	9056
23	26122	2675	2674	5349
14	226309	1544	1539	3083
29	22920	1165	1157	2322
22	268991	837	836	1673
16	345246	560	558	1118
21	40501	285	285	570
30	284	284	284	568
25	175	8	8	16
1	130	2	2	4
13	49620	2	2	4
19	189	1	1	2

domains for CnC infrastructure. For the static lists targeting specific malware families, an explanation for the lack of changes can be that the criminals have moved on to use other malware.

From Table 2 we see that the number of appearances and removals are approximately equal for each list, hence all the list are of approximately constant size. It can also be seen that the numbers of appearances and removals equals the size of the blacklist for BIDs 30 and 31, meaning that the lists potentially have changed completely during the observation period. This is to be compared with BIDs 13 and 16, where only 0.004% and 0.162% of the lists changed. An explanation for this could be that BIDs 13 and 16 addresses domains involved with fraud, for which take-downs actions might be slower.

Figures 2 and 3 show traits that are apparent when maintaining the time dimension. Take for instance BID 13, where both appearances are followed by a removal a few days later, suggesting a relatively quick response, both in addressing the underlying problem and from the list maintainer. For BIDs 14–22 the pattern is instead that we first see some weeks with only appearances, followed by weeks of only removals. Hence one can expect relatively high time lags. Blacklists 29 and 31 show a different pattern, with more evenly distributed changes, indicating more active maintenance. The different types of maliciousness addressed by different lists can also be an explanation.

As future work, we would like to expand our study, with a larger set of blacklists and with a longer period of observation. The large share of static entries

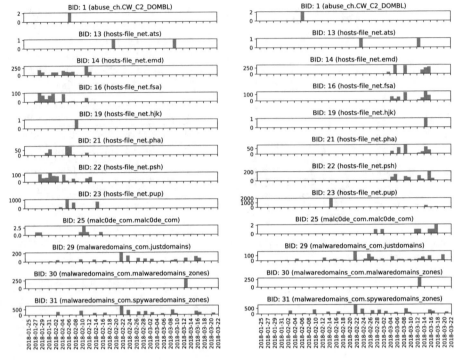

Fig. 2. Appearances per list over time.   **Fig. 3.** Removals per list over time.

might be explained by the lists not seeing sufficiently active maintenance, or by the combination of a limited time span and the nature of the malicious use. If it is the case that domains typically are abused, and therefore blacklisted, for e.g. months then expanding the observation to e.g. a year will improve the analysis. Expanding with more blacklists that target the same types of maliciousness can allow us to determine a typical rate of changes, providing for judging what lists appear to be remarkably above or below the norm. When considering the domains appearing after the epoch, there was too small an intersection between the lists for any meaningful analysis to be made. With a longer period of observation is possible that we can identify interesting relationships between blacklists, such as overlaps, aggregations and typical lag times. One way to expand the set of blacklists is to include query-only blacklists. This would of course raise the issue of how retrievable and query-only blacklists can be compared. Finally, it appears relevant to evaluate blacklists in practice and relate that to our metrics.

6   Conclusion

In this paper we describe how it is a problem to assess the usefulness of blacklists in the absence of a ground truth. We address the problem by proposing a method for analysing blacklists in order to understand how they change over time.

We find large difference among blacklists. Based on this it is apparent that the size of a blacklist alone does not signify its usefulness. Observing blacklists for weeks provides a better picture of which potentially are up to date and therefore useful. Longer running observations and more blacklists are required to understand relationships between blacklists, and we see this as the future direction of this work.

References

1. Antonakakis, M., Perdisci, R., Dagon, D., Lee, W., Feamster, N.: Building a dynamic reputation system for DNS. In: USENIX Security Symposium, pp. 273–290 (2010)
2. Choi, H., Lee, H.: Identifying botnets by capturing group activities in DNS traffic. Comput. Netw. **56**(1), 20–33 (2012)
3. Felegyhazi, M., Kreibich, C., Paxson, V.: On the potential of proactive domain blacklisting. LEET **10**, 6 (2010)
4. Hao, S., Kantchelian, A., Miller, B., Paxson, V., Feamster, N.: Predator: proactive recognition and elimination of domain abuse at time-of-registration. In: Proceedings of the 2016 ACM SIGSAC Conference on Computer and Communications Security, pp. 1568–1579. ACM (2016)
5. Hao, S., Thomas, M., Paxson, V., Feamster, N., Kreibich, C., Grier, C., Hollenbeck, S.: Understanding the domain registration behavior of spammers. In: Proceedings of the 2013 Conference on Internet Measurement Conference, pp. 63–76. ACM (2013)
6. Moura, G.C., Müller, M., Wullink, M., Hesselman, C.: nDEWS: a new domains early warning system for TLDS. In: 2016 IEEE/IFIP Network Operations and Management Symposium (NOMS), pp. 1061–1066. IEEE (2016)
7. Palo Alto Networks, Inc.: Minemeld threat intelligence sharing, 14 March 2018. https://github.com/PaloAltoNetworks/minemeld/wiki
8. Perdisci, R., Corona, I., Giacinto, G.: Early detection of malicious flux networks via large-scale passive DNS traffic analysis. IEEE Trans. Dependable Secur. Comput. **9**(5), 714–726 (2012)
9. Sheng, S., Wardman, B., Warner, G., Cranor, L.F., Hong, J., Zhang, C.: An empirical analysis of phishing blacklists (2009)
10. Stevanovic, M., Pedersen, J.M., D'Alconzo, A., Ruehrup, S., Berger, A.: On the ground truth problem of malicious DNS traffic analysis. Comput. Secur. **55**, 142–158 (2015)
11. Vissers, T., Spooren, J., Agten, P., Jumpertz, D., Janssen, P., Van Wesemael, M., Piessens, F., Joosen, W., Desmet, L.: Exploring the ecosystem of malicious domain registrations in the .eu TLD. In: International Symposium on Research in Attacks, Intrusions, and Defenses, pp. 472–493. Springer (2017)

Switching Network Protocols to Improve Communication Performance in Public Clouds

Sebastian Łaskawiec$^{(\boxtimes)}$, Michał Choraś, and Rafał Kozik

Institute of Telecommunications and Computer Science,
University of Science and Technology UTP Bydgoszcz, Bydgoszcz, Poland
`sebastian.laskawiec@gmail.com`

Abstract. Applications deployed in the cloud may use any of the IP-based network protocols. One of the popular choices is HTTP. However, this protocol is not the best fit for applications with high performance demands. Such applications often take advantage of custom, binary protocols based on TCP transport. Such protocols offer additional capabilities such as asynchronous processing, Mutual TLS for security or compressing the payload. All this features often concentrate on maximizing performance within given constraints. The communication performance plays a crucial role if the application is deployed within more than one cloud. In such scenarios different parts of the application might communicate over the Wide Area Network. Each cloud contains an edge component responsible for handling ingress traffic. Such component is often called, a router.

In this paper we propose two methods for switching network protocols in order to maximize communication performance using a commonly used cloud routers. The first method uses TLS/ALPN extension and aims for secure connection cases. The other method uses HTTP/1.1 Upgrade procedure and can be easily used for transmitting not secured data over the Public Internet.

Keywords: Cloud · Containers · Kubernetes · Router
Communication protocols

1 Introduction

Modern application development trends show that Microservices Architecture deployed on Container-based Cloud quickly gains popularity [11]. Such applications often communicate with each other using HTTP protocol and REST interfaces [13]. Modern container-based cloud solutions, like Kubernetes [3] or OpenShift [17] embrace that model and offer an intuitive way of building large systems using provided building blocks. The most important ones are Services, which represent an internal load balancer and are implemented by IPTables (also known as Netfilter [9]) project. Routing traffic from the outside of the cloud is

© Springer Nature Switzerland AG 2019
M. Choraś and R. S. Choraś (Eds.): IP&C 2018, AISC 892, pp. 224–236, 2019.
https://doi.org/10.1007/978-3-030-03658-4_27

more complicated and can be achieved using at least two ways - exposing a Load Balancer (also known as a Load Balancer Service) or adding a route to a publicly available router (also called an Ingress or a Router; often implemented using HAProxy [2] or Nginx [4]) [12]. Unfortunately, allocating a Load Balancer is often quite expensive, so most of developers focus on using the public router. This enforces HTTP based communication and offers little to none support for custom, binary protocols based on TCP or UDP transports.

In this paper we present two ways of switching network protocols in order to improve communication performance from the outside world to the applications deployed in the cloud. We focus on exploring different ways of using the public router for achieving descent performance at low costs.

This paper is structured as follows: Sect. 2 explains the challenges of using a public router for custom, binary protocols. Section 3 contains discussion on related articles and specifications to work presented in this paper. Section 4 defines requirements and solutions for switching to custom, binary protocols. Experiment results and detailed explanation is presented in Sects. 5 and 6. The last Section contains information about further work.

2   Problem Statement

Most of the on-premise data centers as well as public clouds use a standard routing model, which consists of a reverse proxy, a load balancer and, one or many, application instances (Fig. 1). A standard web application fits perfectly into this model. In most of the cases, an ingress traffic is generated by the end users who use their desktop or mobile devices. Therefore, a reverse proxy is an HTTP oriented component and offers only a limited set of capabilities for other network protocols, such as high performance binary protocols based on TCP (or UDP) transport. Some cloud vendors do not use the latest version of the reverse proxy software and do not fully support HTTP/2 protocol (in case of OpenShift, the full support is scheduled for version 3.11 [7]). This makes achieving descent performance in the cloud environment even more challenging.

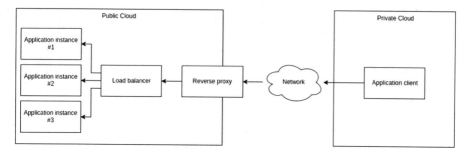

Fig. 1. A standard routing model in modern application deployments

Some of the cloud vendors allow users to allocate an externally reachable L4 Load Balancer (L4 is a transport layer of the OSI Model). Unfortunately, such solutions are often quote expensive considering large scale systems. Table 1 contains a pricing model for an externally reachable Load Balancer provided by major cloud vendors [5,6].

Table 1. Load balancer pricing by the biggest cloud vendors

Cloud provider	Cost
Amazon web services elastic load balancer	0.025 USD per hour and 0.008 USD per GB
Google cloud platform forwarding rule	0.025 USD per hour and 0.008 USD per GB

Adding a new route to an externally reachable reverse proxy is free of charge in most of the cases. Solutions proposed in this paper allow application developers to optimize costs by using a standard reverse proxy and increase the performance of their applications by enabling custom, binary protocols.

3   Related Work

The solution presented in this paper is based on two types of communication protocol switching - HTTP/1.1 Upgrade procedure and TLS/ALPN.

The HTTP/1.1 Upgrade procedure (sometimes called "clear text upgrade") is often used when the client and the server communicate using unencrypted TCP connection. The procedure allows a client application to send an HTTP/1.1 message containing an "Upgrade" header to invite the server to switch the communication protocol reusing the same TCP connection. The server may ignore such a request or send an HTTP/1.1 101 status code (Switching Protocols) and accept one of the proposed protocols by the client. After sending HTTP/1.1 101 message, the server immediately switches to the new protocol reusing the same TCP connection. Figure 2 shows an example of upgrading existing HTTP connection to a custom Hot Rod protocol.

It is also possible to encrypt a connection between a client and a server by using TLS (Transport Layer Security, sometimes called with an older name - SSL). Initiating an encrypted connection requires completing handshake subprotocol. During the handshake, the client and the server may use one of the TLS extension. From this paper's perspective there are two important extensions - SNI (Server Name Indication defined in RFC 3546 [10]) and ALPN (Application-Layer Protocol Negotiation defined in RFC 7301 [1]) The SNI extension transmits an unencrypted "hostname" field, which can be used by the publicly deployed router to make routing decisions. This is a common practice used by the cloud vendors for the reverse proxy (also called router in this

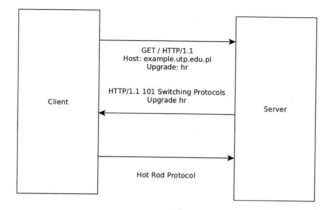

Fig. 2. HTTP/1.1 Upgrade flow

case) implementation. The ALPN extension allows the client and the server to negotiate communication protocol for the given TCP connection. Figure 3 show a simplified diagram of TLS/ALPN communication protocol negotiation.

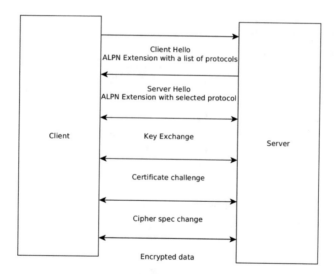

Fig. 3. TLS/ALPN flow

Switching procedure requires also a target protocol. This protocol can be used during the HTTP/1.1 Upgrade procedure as well as for the TLS/ALPN handshake. In this article we discuss possibilities of switching to HTTP/2 protocol (defined in [18]) and Hot Rod Protocol [8] (a custom binary protocol designed for Infinispan and Red Hat Data Grid projects).

From the implementation point of view, this paper continues research started in [14], where we used TLS/SNI extension for designing multi-tenant solution for an open source in-memory data grid solution, called Infinispan. The TLS/SNI "hostname" field was used by two types of routers - a reverse proxy to determine what application should receive data and for the Infinispan Multi-tenant Router that works as a facade separating data containers from each other and allowing multiple users to access their data in secured fashion. This paper reuses TLS/SNI approach to allow the reverse proxy deployed in the cloud to make routing decisions.

Some of the protocols allow using client-side load balancing. The Hot Rod protocol, a binary communication protocol based on TCP transport used by the Infinispan server is one of them. In [12] we proposed using a load balancer per application instance deployed in the cloud (Fig. 4). Such an approach allow to leverage routing optimizations calculated by the client (the client application calculates, which server contains given segment of the data and queries it directly preventing additional network hops). Unfortunately, this solution requires allocating a load balancer per application instance and therefore, might turn out to be very expensive for large deployments.

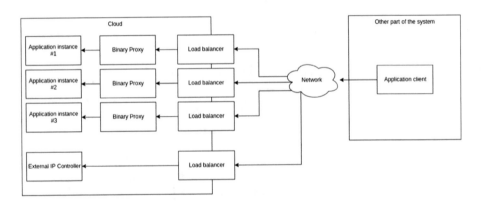

Fig. 4. Load balancer per application instance

4   Requirements and Proposed Solution

Services hosted within the cloud are often allowed to use any communication protocol. Most of the Open Source, container-based clouds use SDN technology (Software Defined Network) to address layer 3 routing (the network layer of the OSI model). Therefore, the only limitation is to use IP-based routing (TCP is often not a requirement and other transport protocols, such as UDP can be used). Handling an ingress traffic in the cloud is often done by using an edge component, a load balancer or a reverse proxy. Some applications need to be hosted in two or more data centers. Such scenarios are often called a "hybrid cloud". A typical example are payments companies, where the transaction processing part is hosted in the public cloud but some sensitive data must be hosted inside a protected data center. In such scenarios it is critical to allow custom protocols communication between the public cloud and other services deployed within the private data center.

Another hard requirement for the solution is encryption. In some cases (especially if a private data center from the previous example hosts sensitive data) it is required to transmit data over wide are network using an TLS encrypted connection.

There are two additional nice-to-have requirements for the proposed solution. The first is the switching procedure, which should be easy to be implemented on the client application side. The reason is to make it less error-prone. The second is to expose all protocols on one TCP port. This lower costs on the service provider side since there's a need only for one route.

The Table 2 contains all gathered requirements with priorities assigned.

Table 2. Solution requirements

Requirement	Priority
Switch to custom protocol	Must have
Possible TLS encryption	Must have
Simple client implementation	Nice to have
Expose all protocols using one TCP port	Nice to have

In this paper we propose two protocol switching mechanisms that fulfill the requirements.

The former is based on TLS/ALPN [1]. During the TLS Handshake subprotocol, the client sends a the custom binary protocol as a first element of TLS/ALPN protocol negotiation list. In most of the cases, the list contains only one protocol supported by the client (one of the most common exceptions is multi-protocol client applications; however, such application are out of the scope of this article). If the server doesn't support the proposed protocol, the connection is terminated.

The latter solution is based on HTTP/1.1 Upgrade procedure [1] and can be used in all the situations where encryption is not necessary. In this solution, the client application needs to be adjusted to send an empty HTTP/1.1 GET request to the root context of the URL with an "upgrade" header. The header is allowed to contain multiple values, however in most of the cases it will contain one element - target binary protocol. The server might ignore such request and in that case, the client terminates the connection.

Both solutions allow switching to a binary protocol and both of them allow to specify multiple protocols in the request (either using TLS/ALPN or HTTP/1.1

Upgrade). In most of the cases TLS/ALPN is handled by OpenSSL library, which supports this extension from version 1.0.2 [16]. In case of Java Programming Language (the implementation for this solution was done using JDK 1.8), the TLS/ALPN is not supported. For the prototype implementation we had to override all classes related SSLContext and SSLEngine and manually handle the ALPN extension. Once JDK 11 becomes new Java LTS release, the solution will be upgraded to use out of the box TLS/ALPN implementation. HTTP/1.1 Upgrade procedure is much more complicated since it needs to be implemented in the application level (rather than transport level). Adding a mechanism to exchange HTTP messages, and then switch to binary protocol is often complicated, therefore it doesn't fulfill the "Simple client implementation" requirement.

The prototype has been implemented using Infinispan Open Source project. Infinispan is an in-memory data store that offers many endpoint protocol implementations, including HTTP/1.1, HTTP/2, Hot Rod, Memcached and Web Sockets. For the proof of concept implementation we picked the most commonly used protocols - HTTP/1.1, HTTP/2 and Hot Rod. Both server and client implementation has been done using Netty framework [15] and can be found in Infinispan ticket system as well as on Github project [19]. The algorithm is based on multi tenancy router implementation [14], which has been modified to perform both TLS/ALPN negotiation and HTTP/1.1 Upgrade procedure. The router implementation manages the communication pipeline (Netty communication is based on Events, which are handled by Handlers within the communication pipeline; It's a "chain of responsibility" design pattern) as well as a full list of supported protocols by the server (which is managed in a routing table). The router requires at least one REST server implementation to be added, because HTTP/1.1 is used as a fallback protocol. A simplified algorithm implementation has been shown in the Fig. 5.

```
Input:  Inbound connection ic
        Routing table rt
          UpgradeHandler uh
          CommunicationPipeline cp
Output: SinglePortUpgradeHandler rd

Protocol p = uh.negotiate(ic)
Handler h = rt.getHandler(p)
if (h == null)
  h = rt.getHandler("HTTP/1.1")
cp.addHandler(h)
```

Fig. 5. Protocol switching implementation

The algorithm is very straightforward and its complexity is proportional to the number of routes in the routing table. Using a Big-O notation it might be written as:

$$T(n) = O(n). \tag{1}$$

5   Experiment Results

All the tests have been done using Java Hot Rod client (an Infinispan client library) and a custom implementation of HTTP/1.1 and HTTP/2 clients (with both TLS/ALPN as well as HTTP/1.1 Upgrade). The testing has been done using Lenovo T470s laptop (Intel(R) Core(TM) i7-7600U CPU, 16 GB of RAM) and OpenShift 3.10.rc0 (Kubernetes 3.10) local cluster. In all the tests there has been a singleton Infinispan Server instance deployed in the cloud and testing harness communicating with the server using different clients and communication paths.

The Table 3 contains performance results for initiating connection and switching to different protocols.

Table 3. Initialize connection results

Negotiation mechanism	Connection type	Target protocol	Iterations	σ	± Error	Result [ms/op]
None	Direct	HTTP/1.1	31	1.048	0.686	2.068
None	OCP Router	HTTP/1.1	31	0.235	0.154	1.087
TLS/ALPN	Direct	HTTP/2	31	0.811	0.531	5.063
TLS/ALPN	OCP Router	HTTP/2	31	2.343	1.535	6.576
HTTP/1.1 Upgrade	Direct	HTTP/2	31	1.320	0.864	3.310
HTTP/1.1 Upgrade	OCP Router	HTTP/2	31	2.470	1.617	4.464
TLS/ALPN	Direct	Hot Rod	31	1.683	1.102	9.742
TLS/ALPN	OCP Router	Hot Rod	31	1.988	1.302	10.401
HTTP/1.1 Upgrade	Direct	Hot Rod	31	1.630	1.067	5.389
HTTP/1.1 Upgrade	OCP Router	Hot Rod	31	15.457	10.122	8.594

Tables 4 and 5 contain performance results for sending 360 and 36 byte entries (as a key/value pairs) into the server. The results have been visualized in Figs. 6, 7 and Fig. 8.

Table 4. Uploading 360 bytes to the server results

Negotiation mechanism	Connection type	Target protocol	Iterations	σ	± Error	Result [ms/op]
None	Direct	HTTP/1.1	31	0.504	0.330	0.472
None	OCP Router	HTTP/1.1	31	1.192	0.781	1.315
TLS/ALPN	Direct	HTTP/2	31	0.431	0.282	1.577
TLS/ALPN	OCP Router	HTTP/2	31	0.732	0.480	2.149
HTTP/1.1 Upgrade	Direct	HTTP/2	31	0.120	0.078	1.048
HTTP/1.1 Upgrade	OCP Router	HTTP/2	31	0.119	0.078	1.156
TLS/ALPN	Direct	Hot Rod	31	0.060	0.039	0.269
TLS/ALPN	OCP Router	Hot Rod	31	0.074	0.048	0.331
HTTP/1.1 Upgrade	Direct	Hot Rod	31	0.056	0.037	0.193
HTTP/1.1 Upgrade	OCP Router	Hot Rod	31	0.078	0.051	0.255

Table 5. Uploading 36 bytes to the server results

Negotiation mechanism	Connection type	Target protocol	Iterations	σ	± Error	Result [ms/op]
None	Direct	HTTP/1.1	31	0.515	0.337	0.426
None	OCP Router	HTTP/1.1	31	1.163	0.761	2.310
TLS/ALPN	Direct	HTTP/2	31	0.299	0.196	0.754
TLS/ALPN	OCP Router	HTTP/2	31	0.153	0.100	0.649
HTTP/1.1 Upgrade	Direct	HTTP/2	31	0.094	0.062	0.267
HTTP/1.1 Upgrade	OCP Router	HTTP/2	31	0.074	0.052	0.302
TLS/ALPN	Direct	Hot Rod	31	0.094	0.062	0.267
TLS/ALPN	OCP Router	Hot Rod	31	0.074	0.052	0.302
HTTP/1.1 Upgrade	Direct	Hot Rod	31	0.100	0.065	0.231
HTTP/1.1 Upgrade	OCP Router	Hot Rod	31	0.061	0.040	0.243

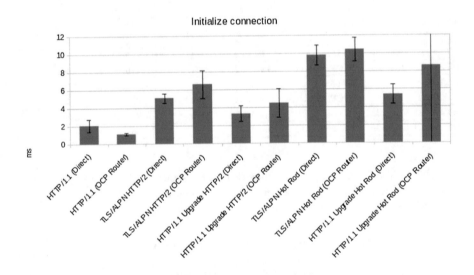

Fig. 6. Initialize connection results

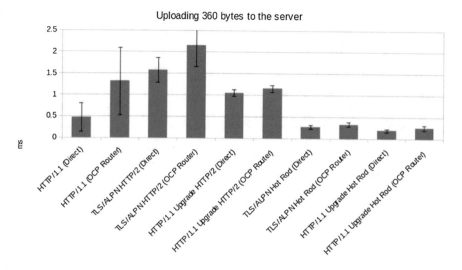

Fig. 7. Uploading 360 bytes to the server results

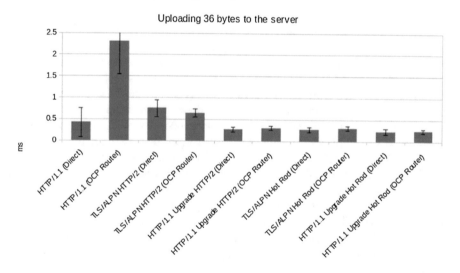

Fig. 8. Uploading 36 bytes to the server results

6   Conclusions

Selecting proper network protocol transmitting data between services deployed in different clouds is not a trivial topic. Each application and the each development team has its own requirements for the technology used in the project. Some of the commonly used requirements have been gathered in Table 2. Most of the languages and out of the box libraries provide support for HTTP/1.1 protocol (some of them offer also seamless upgrade to HTTP/2 but at the moment of

Table 6. Use cases and recommendations

Use case	Key deciding factors	Protocol switch mechanism	Target protocol
Transmitting sensitive data at high speed	Performance encryption	TLS/ALPN	Custom, binary protocol
Transmitting public at high speed	Performance	HTTP/1.1 Upgrade	Custom, binary protocol
Occasionally send short heartbeats	Fast connection initialization	HTTP/1.1	HTTP/1.1
Transmitting sensitive data without new dependencies	No new dependencies encryption	TLS/ALPN	HTTP/2

writing this article most of the Java clients lack support of it), which was one of the major contributing factor for its popularity among application and library developers. However, in some scenarios, simplicity is not a deciding factor. Performance or asynchronous processing allows application to increase performance and lower operating costs at the same time (very often those two things are tied together - better performance allows to decrease the number of application replicas and lower the costs). In order to reduce costs even further, it is highly advisable to use a publicly available cloud router instead of dedicated, often quite expensive, application load balancer. Such routers are often designed with HTTP/1.1 and TLS in mind. Supporting custom protocols can be achieved by encrypting the traffic or using HTTP/1.1 Upgrade procedure. During the experiment we benchmarked the time for initiating connection to an in-memory data store - Infinispan (Table 3 and Fig. 6). The first two rows (no switching mechanism and HTTP/1.1 used as a communication protocol) has been used as a baseline. Introducing a public router between the testing harness and the server (connection type equal to "OCP Router") made the connection initialization roughly 2 ms slower than without it. Each pair of results present similar pattern - using HTTP/1.1 Upgrade procedure is slightly faster than TLS/ALPN. The worst result was observed for the Hot Rod Protocol, which performs its own handshake procedure and exchanges server information during connection initialization. As the results show, exchanging this information takes, quite a long time (comparing to the other measured protocols). The next important benchmark is inserting 36 and 360-byte keys and values into the in-memory store. As expected, in all cases using unencrypted connections was faster then connections with TLS. The difference get bigger with larger payloads. This is also expected, since encrypted payloads are larger than unencrypted once. The custom protocol used for testing performed more than twice faster than HTTP/1.1.

HTTP/2 performance was slightly lower than Hot Rod's but it was still faster than HTTP/1.1. At this point, it is also worth to mention, that the testing procedure was synchronous (the testing harness sent a request to the server and waited for response). Both HTTP/2 and Hot Rod protocols support asynchronous processing and using it would make the performance gap even larger. The results clearly show that using custom, binary protocols for inter-cloud communication is not only possible, but often highly advisable when performance is one of the key factors. Even though initiating connection usually takes a bit longer than with HTTP/1.1, the overall communication throughput is much higher. In our tests, the Hot Rod protocol offered the best performance with reasonably low standard variation. However, this requires application developers to introduce another dependency to their project. In some environments, this might be not feasible. In such cases, the HTTP/2 protocol offers slightly lower results (compared to Hot Rod) but it is often supported by out of the box libraries in programming languages. We consider HTTP/2 protocol as a middle ground between HTTP/1.1 text protocol and custom, binary protocols. The Table 6 shows some of the use cases with suggested protocols and switching solutions.

7 Further Work

Both research and implementation has been done using Infinispan Open Source project. At the moment of writing this article, the server implementation has already been merged into the project, whereas the client part is still waiting to be reviewed by the Infinispan Team. In the near future we plan to expand this work to other server protocols such as Web Sockets or Memcached and compare the results. Another goal is to create a multi-protocol client implementation which can switch communication protocols on demand. Such an implementation allows to address very interesting use cases that require different connection characteristics to perform different tasks. An example of such behavior is transmitting a large amount of encrypted traffic along with small portions of public data. In that case the client could use a TLS/ALPN encrypted Hot Rod connection along with HTTP/1.1 protocol.

References

1. RFC 7230 - Hypertext Transfer Protocol (HTTP/1.1): Message Syntax and Routing (2014). https://tools.ietf.org/html/rfc7230#page-57
2. HAProxy web page (2017). http://www.haproxy.org/
3. Kubernetes web page (2017). https://kubernetes.io/
4. Nginx web page (2017). https://nginx.org/en/
5. AWS pricing (2018). https://aws.amazon.com/elasticloadbalancing/pricing/
6. Google Cloud pricing (2018). https://cloud.google.com/pricing/
7. Implement router http/2 support ticket (2018). https://trello.com/c/qzvlzuyx/27-3-implement-router-http-2-support-terminating-at-the-router-router
8. Infinispan 9.3 User Guide (2018). http://infinispan.org/docs/stable/user_guide/user_guide.html#hot_rod_protocol_2_8

9. The netfilter.org project web page (2018). https://www.netfilter.org/
10. 3546, R.: RFC3546 - Transport Layer Security (TLS) Extensions (2003). https://tools.ietf.org/html/rfc3546
11. Laskawiec, S.: The Evolution of Java Based Software Architectures. C. Int. Publ. J. Cloud Comput. Res. 2(1), 1–17 (2016)
12. Laskawiec, S., Choraś, M.: New Solutions for exposing Clustered Applications deployed in the cloud (2018). https://doi.org/10.1007/s10586-018-2850-3
13. Fielding, R.T., Software, D., Taylor, R.N.: Principled Design of the Modern Web Architecture. Technical report 2 (2000). https://www.ics.uci.edu/~taylor/documents/2002-REST-TOIT.pdf
14. Laskawiec, S., Choraś, M.: Considering service name indication for multi-tenancy routing in cloud environments. In: Advances in Intelligent Systems and Computing, vol. 525, pp. 271–278 (2017)
15. Netty: Netty web page (2018). https://netty.io/
16. OpenSSL: Openssl 1.0.2 release notes (2018). https://www.openssl.org/news/openssl-1.0.2-notes.html
17. Red Hat, I.: OpenShift: Container Application Platform by Red Hat, Built on Docker and Kubernetes (2017). https://www.openshift.com/
18. RFC7540: RFC 7540 - Hypertext Transfer Protocol Version 2 (HTTP/2) (2015). https://tools.ietf.org/html/rfc7540
19. Singleportjira: [ISPN-8756] Implement Single Port - JBoss Issue Tracker (2018). https://issues.jboss.org/browse/ISPN-8756

Interference-Aware Virtual Machine Placement: A Survey

Sabrine Amri[1(✉)], Zaki Brahmi[1], Rocío Pérez de Prado[2],
Sebastian García-Galán[2], Jose Enrique Muñoz-Expósito[2],
and Adam Marchewka[3]

[1] Sousse University, Sousse, Tunisia
sabrine3amri@gmail.com, zakibrahmi@gmail.com
[2] Telecommunication Engineering Department CCT Linares, University of Jaén,
Jaén, Spain
{rperez,sgalan,jemunoz}@ujaen.es
[3] UTP University of Science and Technology in Bydgoszcz, Bydgoszcz, Poland

Abstract. In order to maintain energy efficiency and optimize resource
utilization in cloud datacenters, cloud providers adopt virtualization
technologies as well as server consolidation. Virtualization is the means
used to achieve multi-tenant environments by creating many virtual
machines (VMs) on the same physical machine (PM) in a way that they
share same physical resources (e.g. CPU, disk, memory, network I/O,
etc.). Server consolidation consists of placing as much as possible VMs
on as less as possible PMs, in the aim of maximizing idle servers and
then minimize the energy consumption. However, this vision of server
consolidation is not as simple as it seems to be. Hence, it is crucial to be
aware of its emerging concerns, such as severe performance degradation
problem when placing particular VMs on the same PM. Furthermore,
the Virtual Machine Placement (VMP) is one of the most challenging
problems in cloud environments management and it's been studied from
various perspectives. In this paper, we are going to propose a VMP taxonomy in order to understand the various aspects that researchers consider
while defining their VMP approaches. We will also survey the most relevant interference-aware virtual machine placement literature and then
we shall give a comparative study between them.

1 Introduction

Cloud computing environments are multi-tenant based environments. Hence,
many virtual machines (VMs) can be allocated in the same physical machine
(PM) thanks to the virtualization technologies. The main idea behind these
technologies is to maximize resource utilization per physical machine using server
consolidation in a way that the under-used server can be shut down and therefore lower the energy consumption. Despite its advantages such as security isolation [5] which prevents malicious data access and fault isolation [6] that guaran-
tees convenient execution of VMs while another VM fails, virtualization technologies still suffer from the lack of effective performance isolation among co-hosted

© Springer Nature Switzerland AG 2019
M. Choraś and R. S. Choraś (Eds.): IP&C 2018, AISC 892, pp. 237–244, 2019.
https://doi.org/10.1007/978-3-030-03658-4_28

VMs [3]. Thus, the performance of one VM can be affected by the behavior of another adversely one, both sharing same system's physical resources. Furthermore, the VM interference problem as it was studied and demonstrated [1–3] is the severe performance degradation that occurs within co-located VMs due to contention over shared resources. In some cases, it may even lead a virtual machine to stop responding [8–10]. Therefore, VM interference problem should be well considered when allocating VMs across PMs.

The rest of this paper is organized as follows, Sect. 2 defines the VMP problem and emphasizes a taxonomy of VMP problem related works. Section 3 reviews the VMP approaches. The comparative study of the surveyed papers is reveled in Sect. 4. Finally, Sect. 5 concludes the paper and discusses the comparative study.

2   Proposed Taxonomy of Virtual Machine Placement

VM Placement. VM placement is considered as an NP-hard bin packing problem [7] with differently sized bins. To solve it, many researches have been performed to define their own policies with adopting many aspects that will be discussed in the next section. Formally the VM Placement Problem VMPP is described by the triple:

$$VMPP = \prec PM, VM, aff \succ$$

- $PM = \{p_1, p_2, ..., p_m\}$ the set of m PMs. Each PM $p_j \in PM(j \in [1..m])$ possesses a d-dimensional Server Capacity Vector SCV_{p_j}. $SCV_{p_j} = \langle C_{p_j}^1, C_{p_j}^2, ..., C_{p_j}^d \rangle$ where $C_{p_j}^k$ denotes the total capacity of resource K; $K \in [1..d]$
- $VM = \{v_1, v_2, ..., v_n\}$ the set of n virtual machines. Each virtual machine $v_i \in VM$ possesses a d-dimensional Resource Demand Vector RDV_{v_i}. $RDV_{v_i} = \langle D_{v_i}^1, D_{v_i}^2, ..., D_{v_i}^d \rangle$, where $D_{v_i}^k$ denotes the total capacity of resource. The total PM utilization is described as the sum of VM demands and it possesses a d-dimensional PM Utilization Vector SUV_{p_j}. $SUV_{p_j} = \langle U_{p_j}^1, U_{p_j}^2, ..., U_{p_j}^d \rangle$.
  $U_{p_j}^k = \sum_{i=1}^n D_{v_i}^k \forall v_i \in VM_{p_j}$ VM_{p_j} denotes the VM set that belongs to PM_{pj}
- $f : PM \times VMs \Longrightarrow \{\langle v_i - p_i \rangle\}; v_i \in VMs , p_i \in PM$

2.1   Proposed Taxonomy

Based on studying the most relevant VMP existing research articles, we came up with classifying them according to two levels: (1) the studied aspect and (2) the used approach as shown in Fig. 1. In this paper, we are going to review research papers dealing with virtual machine placement problem from the Energy-aware aspect regarding the inter-VM performance interference issue when placing VMs on PMs.

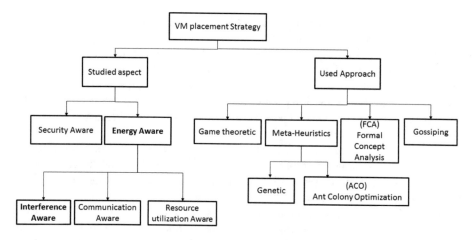

Fig. 1. Taxonomy of the VMP approaches

3 Reviewed Approaches Regarding Interference-Aware VMP

The VMP problem is being widely studied from various perspectives as we've discussed in the previous section. One of the major VMP related aspects is interference-aware. In this section, we emphasize existing researchers' proposed approaches to map VMs into the available PMs regarding energy efficiency optimization through building VMP models and algorithms aiming to mitigate the inter-VM performance interference.

On Jersak et al. [11], authors define an algorithm for server consolidation to map VMs into a minimal number of PMs using an adjustable, user defined, VM interference threshold. In a way that users can have a better tradeoff between amount of needed resources and performance they want to reach, by keeping interference level under a given threshold that they define. Their algorithm focuses on minimizing used PM number by using a virtualization interference model and classifying VMs according to their resource utilization profile.

In Shim [12], introduce a static VM consolidation algorithms to reduce energy consumption with considering the inter-VM interference in a way that the VM consolidation algorithm do not incur too much jobs performance degradation between co-located VMs. He considers PMs with homogeneous hardware capacity and he compares interference between jobs running on the same PM core to when jobs are running on different cores on one PM. The author states that jobs running on different cores interfere less with each other than those running on the same core.

Moreno et al. [13], define an Interference-Aware Allocation module that takes into account the workload heterogeneity and uses decision-making techniques to select the best workload host according to its internal interference level. The allocation is based on specific workload types to improve both performance of running tasks and energy-efficiency.

Kim et al. [14], propose a VM consolidation algorithm based on cache interference-sensitivity and interference-intensity VMs. Interference intensity measures how much an application will hurt others by taking away their resources and the interference sensitivity measures how much an application will suffer from others. Their idea was to co-locate highly interference-intensive VMs with less interference-sensitive ones.

Jin et al. [15], introduce a VM placement model based on cache interference and employing commonly reuse distance analysis to characterize HPC cloud applications. VMs are dispatched to distinct cores based on co-located application's cache behavior and cache interference intensities.

Lin et al. [16], propose a heuristic-based algorithm to solve the Interference-aware VM placement problem. They define a placement strategy based on improving application's QoS with maximizing cloud provider's profit. In Caglar et al. [17], authors define a real-time performance interference-aware VM placement model, using machine learning techniques that classify VMs based on their historic mean CPU and memory usage. To allocate the classified VMs they extract patterns that provide the lowest performance interference level on the specified host machines.

All these works study the interference-aware VMP problem in the aim of maximizing energy efficiency and minimizing inter-VM performance interference, which are two opposite goals to accomplish while consolidating servers and allocating VMs to existing PMs.

4   Comparative Study

In this section, we establish a comparative study of the previously exhibited VMP approaches based on a set of criteria that we define in the next subsection. This comparative study may clarify the similarities and the mismatches between the most relevant virtual machine placement problem approaches.

4.1   Used Criteria

Our comparative study is based on the following criteria:

- **Used resources**: VM defined profiles in each VMP approach. It consists on resources that present intensive utilization to experience interference (e.g. CPU, RAM, Disk I/O, etc.).
- **Interference metrics**: represents the used parameters to compute interference. Precisely, parameters to measure the performance degradation.
- **Placement metrics**: identifies the adopted parameters to verify by the placement module while mapping VMs to PMs. Placement goal: defines the main objective of server consolidation while placing the VMs (e.g. minimize active PMs number, Minimize interference, etc.).
- **Placement approach**: specifies the authors adopted approach to resolve the VMP problem.

Table 1. Comparative study of interference-aware VMP approaches

Approach	Used resources	Interference	Placement metrics	Placement goal	Placement approach
[11]	CPU, RAM, Disk I/O	CPU execution time percentage of increase RAM and Disk I/O, throughput percentage decrease	Interference threshold	Minimize PM number	Heuristics (First Fit, Best Fit, Worst Fit)
[12]	CPU, Memory, Disk I/O	-slowdown(job1@job2): ratio ("completion time of job1 when job2 is running on the Same PM" To "completion time of job1 when running alone on the PM")	Ratio of idle PMs, Ratio of SLA violation	Allocate interfering jobs on different cores of the same PM. Minimize energy consumption and, Reduce performance degradation between co-located jobs	Heuristics (Modified Best Fit)
[13]	CPU, Memory	QoS in term of throughput, latency, and Response time	Workload heterogeneity	Select the best workload host according to its internal interference level	Decision-making techniques
[14]	CPU cache(LLC), Memory bus	LLC misses and references	Interference intensity and Interference sensitivity	Minimize average performance degradation ratio of all applications	–
[15]	CPU cache (LLC)	Distance analysis	Cache interference intensity and Cache pollution	Schedule VMs based on Identifying application's cache intensity	Classification
[15]	CPU cache (LLC)	Distance analysis	Cache interference intensity and Cache pollution	Schedule VMs based on Identifying application's cache intensity	Classification
[16]	Network I/O	QoS violation: service time of network I/O request	Resource demand of a VM, Application QoS, VM interference	Reduce VM interference, Fully exploit resource capacities of PMs, Satisfy applications QoS requirements, Maximize cloud provider's profit	Integer Linear programming, Polynomial-time Heuristic
[17]	CPU, Memory	Cycles Per Instruction (CPI): response time	Lowest performance interference level	Online VM placement strategy	Heuristics, Machine Learning

4.2   Emerging Comparison

Table 1 shows the comparative study for works, dealing with the performance interference-aware VMP problem, that we consider most relevant and outline the reported literature. In what follows we build our comparative overview based on the set of criteria depicted in the previous section. The comparative study shows that none of the surveyed approaches considers VMP problem from different levels of the cloud computing environments (i.e. VM level, PM level and Application level).

5   Conclusion and Discussion

Server consolidation is considered as one of the main features of virtualization in modern Data Centers. However, it is also considered as an NP-hard bin packing problem [7] which makes it one of the most controversial research challenges in the cloud computing environments [18]. In this paper, we defined our taxonomy of existing works related to the VPM problem. We then picked one of the most studied VMP energy-based aspects to survey, which is the performance interference problem. Finally, we've carefully chosen and read the most relevant literature to this topic and we've established a comparative study between them.

Our study showed that despite their focus on the virtual machine interference problem while minimizing energy consumption when placing VMs on PMs in cloud Data Centers, most of the VMP approaches did not consider all levels of cloud computing environment which are the VM, PM and Application levels. Hence, each work studies the interference problem based only one particular level without seriously considering the rest of other levels. Therefore, in order to define an efficient VMP solution we must consider all levels of a cloud computing environment (i.e. VM, PM and application levels), we must also study the diverse resource-bound workloads (i.e. CPU, Disk, Network I/O, Memory intensive workloads) to balance resource usage and mitigate resource contention between VMs. Another important aspect toward an efficient network-bound resource utilization is to consider inter-VM communication since some research works on network communication prove that communication between VMs should be seriously considered to save energy communication cost between network elements within a data center [19].

References

1. Hamilton, J.: Cooperative expendable micro-slice servers (CEMS): low cost, low power servers for internet-scale services. In: Conference on Innovative Data Systems Research (CIDR 2009) (January 2009)
2. Nathuji, R., Kansal, A., Ghaffarkhah, A.: Q-clouds: managing performance interference effects for qos-aware clouds. In: Proceedings of the 5th European Conference on Computer systems. ACM (2010)

3. Koh, Y., Knauerhase, R., Brett, P., Bowman, M., Wen, Z., Pu, C.: An analysis of performance interference effects in virtual environments. In: IEEE International Symposium on Performance Analysis of Systems and Software, ISPASS 2007, pp. 200–209. IEEE (2007)

4. Pu, X., Liu, L., Mei, Y., Sivathanu, S., Koh, Y., Pu, C.: Understanding performance interference of I/O workload in virtualized cloud environments. In: 2010 IEEE 3rd International Conference on Cloud Computing (CLOUD), pp. 51–58. IEEE (2010)

5. Murray, D.G., Milos, G., Hand, S.: Improving Xen security through disaggregation. In: Proceedings of the Fourth ACM SIGPLAN/SIGOPS International Conference on Virtual Execution Environments. ACM (2008)

6. Nagarajan, A.B., Mueller, F., Engelmann, C., Scott, S.L.: Proactive fault tolerance for HPC with Xen virtualization. In: Proceedings of the 21st Annual International Conference on Supercomputing, pp. 23–32 (2007)

7. Chekuri, C., Khanna, S.: On multidimensional packing problems. SIAM J. Comput. **33**(4), 837–851 (2004)

8. Matthews, J.N., Hu, W., Hapuarachchi, M., Deshane, T., Dimatos, D., Hamilton, G., McCabe, M., Owens, J.: Quantifying the performance isolation properties of virtualization systems. In: Proceedings of the 2007 Workshop on Experimental Computer Science, p. 6. ACM (2007)

9. Padala, P., Zhu, X., Wang, Z., Singhal, S., Shin, K. G.: Performance evaluation of virtualization technologies for server consolidation. HP Labs Technical Report (2007)

10. Xavier, M.G., Neves, M.V., Rossi, F.D., Ferreto, T.C., Lange, T., De Rose, C.A.: Performance evaluation of container-based virtualization for high performance computing environments. In: 2013 21st Euromicro International Conference on Parallel, Distributed and Network-Based Processing (PDP), pp. 233–240. IEEE (2013)

11. Jersak, L.C., Ferreto, T.: Performance-aware server consolidation with adjustable interference levels. In: Proceedings of the 31st Annual ACM Symposium on Applied Computing, pp. 420–425. ACM (2016)

12. Shim, Y.-C.: Inter-VM performance interference aware static VM consolidation algorithms for cloud-based data centers. Recent Adv. Electr. Eng. 18 (2015)

13. Moreno, I.S., Yang, R., Xu, J., Wo, T.: Improved energy-efficiency in cloud data-centers with interference-aware virtual machine placement. In: 2013 IEEE Eleventh International Symposium on Autonomous Decentralized Systems (ISADS), pp. 1–8. IEEE (2013)

14. Kim, S.G., Eom, H., Yeom, H.Y.: Virtual machine consolidation based on interference modeling. J. Supercomput. **66**(3), 1489–1506 (2013)

15. Jin, H., Qin, H., Wu, S., Guo, X.: CCAP: a cache contention-aware virtual machine placement approach for HPC cloud. Int. J. Parallel Program. **43**(3), 403–420 (2015)

16. Lin, J.-W., Chen, C.-H.: Interference-aware virtual machine placement in cloud computing systems. In: 2012 International Conference on Computer and Information Science (ICCIS), vol. 2, pp. 598–603. IEEE (2012)

17. Caglar, F., Shekhar, S., Gokhale, A.: Towards a performance interference-aware virtual machine placement strategy for supporting soft real-time applications in the cloud. In: Proceedings of the 3rd IEEE International Workshop on Real-time and distributed Computing in Emerging Applications, (Co-located with 35th IEEE RTSS), Italy, 15–20 2011
18. Zhang, Q., Cheng, L., Boutaba, R.: Cloud computing: state-of-the-art and research challenges. J. Internet Serv. Appl. **1**(1), 7–18 (2010)
19. Brahmi, Z., Hassen, F.B.: Communication-aware VM consolidation based on formal concept analysis. In: Proceedings of AICCSA 2016, pp. 1–8 (2016)

System Architecture for Real-Time Comparison of Audio Streams for Broadcast Supervision

Jakub Stankowski$^{(\boxtimes)}$, Mateusz Lorkiewicz, and Krzysztof Klimaszewski

Chair of Multimedia Telecommunications and Microelectronics,
Poznań University of Technology, Poznań, Poland
{jstankowski,mlorkiewicz,kklima}@multimedia.edu.pl

Abstract. The paper describes a system for supervision of broadcast streams in terms of correct audio stream assignment to different programmes. The system utilizes a fast method of audio data comparison and is able to perform in real time, even for a significant number of compared data streams and even on performance limited platforms such as Raspberry Pi. The architecture of the system is modular, therefore future extensions can be easily added and the functionality of the system can be expanded according to requirements.

1 Introduction

The idea behind the work described in this paper comes from one of the Polish broadcasters - Telewizja Puls, who needed a way to automatically test, whether the transport streams assembled at the headend contain the correct audio tracks and that audio data is actually present at all for a given programme forwarded to the viewers. The question of having correct audio streams being sent to the viewers is indisputably one of the main factors influencing the quality of experience of viewers. Therefore it is of the utmost importance to ensure that no errors are made in the audio streams assignment. The errors that are supposed to be detected by the software are the following: missing audio stream, wrong audio stream (i.e. audio stream from a different programme), too large difference of audio level between stereo channels. Also the delay between reference stream and the correctly assigned audio data should be measured.

The answer to the needs was a system developed for the purpose of audio stream comparison and verification. The nature of the application demands that the system works in real time and is able to recognize errors as soon as possible to shorten the time when incorrect data is forwarded to the viewers. Due to the requirement of the real time operation, the system cannot employ the state of the art algorithms used for the audio comparison, since the most frequently used systems utilizing fingerprints operate on large chunks of audio data, at least of several seconds in duration. This is largely too long for the purpose of the application and requires a significant amount of processing power. A faster method of

M. Choraś and R. S. Choraś (Eds.): IP&C 2018, AISC 892, pp. 245–252, 2019.
https://doi.org/10.1007/978-3-030-03658-4_29

audio data comparison is required. It needs to meet several requirements, stemming from the nature of the application. Among other requirements, it needs to be immune to transcoding of audio data (i.e. recoding of the compressed audio stream at the headend to a different bitrate or a different format) either in a heterogeneous (using different codecs) or homogeneous (using the same codec for decoding and encoding, only with different settings) manner. It also needs to be prepared for different delays caused by the processing of the data. Such delays, providing they stay below certain threshold, should not be regarded as errors.

In order to be able to process the data in real time, a simple method is required, that does need long buffers for proper operation. The most obvious way of comparing two audio data streams – comparing or cross correlating the consecutive samples of audio, does not work in this situation, since the audio streams are usually recoded using lossy algorithms. Those algorithm preserve the subjectively perceived content of audio stream, but do not preserve the actual sample values of audio data. Therefore, a dedicated algorithm was prepared for the purpose of the system. The details of the algorithm implemented in the system are described in [6], but a short summary is given further in the text.

In order for the aforementioned algorithm to be used in a working system, several additional modules need to be implemented. They provide data for the algorithm by extracting and synchronizing the data from the input streams and signal the current state of the audio streams under consideration. The modules need to be optimized so that they perform their tasks efficiently, in order to be able to take advantage of the speed of the comparison algorithm. They should not cause any excessive delays. All the necessary modules, as well as their interfaces, are described in this paper.

In this paper we show the actual implementation of the audio stream comparison algorithm in a real life system and assess its overall performance. The system performance is evaluated using two different platforms. One of the platforms is a personal computer based on powerful Intel i7 processor, as it was the platform originally used at the headend. However, since the system proved to consume low amount of the resources, a decision was made to port it to a single board computer. We decided to use a very common board Raspberry Pi 3 model B, produced and marketed by Raspberry Pi Foundation [3]. All the evaluations are performed using a specific implementation for the specific hardware platform. Only minor changes were required in the source code in order to port the system to Raspberry Pi (i.e. removal of x86-64 specific intrinsics).

2   Proposed System Architecture Overview

The system for real-time supervision of broadcast audio content has been designed in a modular way, in order to prepare it for being ported to different platforms. This way it is also relatively easy to make any changes or additions that might be needed or desired in future.

What is important for the performance of the system, in order to avoid any bottlenecks and possible locks, each stage operates in separate thread and is separated from each other by thread-safe FIFO (first in, first out) queues.

The main general architecture of a system has been shown on Fig. 1. The system contains a number of receiver stages (each for every input transport stream), a central synchronizer stage and number of analyser stages (one for every compared pair of audio streams).

In proposed architecture, one of the input streams is treated as a reference one and all the other streams are compared against the reference one. Therefore, for N input streams, N receiver stages and N-1 analyser stages are required.

At the beginning of the research, it was stated that the system should be able to receive input data streams in the form of transport streams provided over IP network (TS over IP) in a form of Single Program Transport Stream (SPTS).

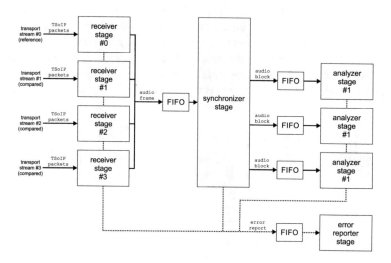

Fig. 1. Proposed system overview (example for 4 input streams)

3 Receiver Stage

A general purpose of the receiver stage is data acquisition and extraction of proper audio stream. It is presented on Fig. 2. First step is MPEG-2 Transport Stream (TS) [1] receiving. Module initiates a connection to Multicast server that provides the required TS. As already stated, in our application a Single Program Transport Stream (SPTS) was used. Data are received in the form of UDP packets, then acquired data are separated to TS packets.

Next step is TS demultiplexation process. TS packets, with Packet Identifier (PID) different than the one defined in configuration, are rejected. Then audio

Elementary Stream Packets are reconstructed and stream type is checked. If read type is different than audio format the error is reported.

Otherwise, the audio data packet extracted from Packetized Elementary Stream (PES) is forwarded to audio decoder. In our implementation MPEG-1 Layer 2 [2] decoder was used, but any other popular encoding format can be used, thanks to the modular structure of the system. To provide further synchronization the Presentation Time Stamp (PTS) is attached to decoded data. Depending on synchronization mode, PTS is received based on System Clock Reference or estimated from UDP packet arrival time. Second mode is much less accurate, due to the fact that single UDP packet can contain data from more than one frame, but may be the only option in case of supervising streams from a different multiplexer for example controlling stream from external operator. Decoded audio data with PTS are then sent to Synchronizer stage.

Presented functionality is provided by using libav library from ffmpeg [4] project. This programming library is used to perform Multicast reception, demultiplexing and decoding functionality. What is more, it allows to use other transport streams and coding formats [4], but in our application these are fixed to previously presented. For every audio stream that is supervised, a separate instance of this stage is created. Instance number 0 always provides reference stream. Data received from all instances are put into one FIFO interface to next stage.

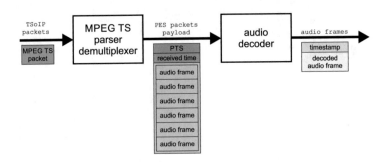

Fig. 2. Receiver stage operation

4   Synchronizer Stage

The synchronizer is responsible for buffering of received audio frames and matching of corresponding frames according to timestamp.

Data from input queue is stored in individual buffers, one for each input stream. The synchronizer keeps track of timestamp of the oldest and the latest packet in each buffer and timestamp of last frame sent to output queue. As soon as matching data is available, the synchronizer merges required number of input frames into longer block and sends it to the analyser stage (Fig. 3).

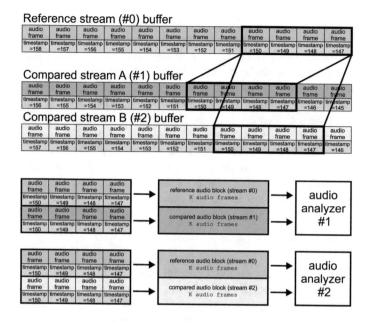

Fig. 3. Example of synchronizer stage operation for two compared streams

The synchronizer has many features which prevent the comparison of false or invalid data. In case of error in receiver stage the synchronizer has the ability to flush one or all audio frame buffers. This allows to avoid comparison against broken or invalid data. The error could be related to lost Transport Stream packets, invalid packets, data error, decoding error and number of different conditions. The information about error in receiver stage is passed by the same queue which is used to pass decoded frames.

The special case which has to be handled in synchronizer stage is so called "PCR rollover" event. PCR (Program Clock Reference) is transmitted as a tick of 90 kHz clock and the transport stream syntax provides field for 33 bits of tick counter. Due to limited length of PCR field, the PCR value has to be reset to zero after ~26 h of operation. The reference clock reset event also breaks the synchronization of stream.

5   Analyser Stage

This stage is used for detecting all problems related to audio data. First step, after obtaining synchronized data from reference and compared streams, is calculating the power of both stereo channels of given signals. Then two conditions are checked: is power of any audio data channel is lower than threshold and if power difference between channels in both audio streams is bigger than defined value.

The purpose of the first condition is to prevent situation, when signal contains "silence" which may cause improper results of the comparison algorithm. Second one may be very useful to provide proper quality of service given by program studio to user of TV system. If any of the described situations occurs, the further calculations are stopped. What is more, if any condition is true for a number of times defined in system configuration, the error is reported.

Next step is comparing audio streams using authors' algorithm [6]. The envelope of reference and compared data is calculated. Then the special descriptor value is determined, as described in [6]. If descriptor value is lower than a specifc threshold, compared audio streams are marked as being different. In other case, the delay between streams is measured. To avoid the situations when the error is reported to the user for correctly assigned and processed stream, the error reporting is initiated only after the error is detected a few times in a row, as specified in the configuration file of the system.

6   Error Reporter Stage

To provide error reporting to system administrator, the error reporter stage is used. Reports from every stage instances are acquired using FIFO interface. Every report contains information about type of the error, time of its appearance, reporting stage name and stream ID. Based on error type, proper information for system administrator is created. System statement contains all error report information with short error description and its code.

The statement can be delivered to administrator in two ways. The most basic is creating dedicated log file in defined directory. Other method is sending information to remote logging system i.e. rsyslog [5].

Presented stage can be used for other purposes than error reporting. For example separate instance can perform performance monitor role. Change of module functionality requires connection to a different report interface and changing report messages.

7   Inter-module Communication

Each processing unit (like receiver, synchronizer analyzer, etc.) is implemented as a separate pipeline stage. Each pipeline stage has its own dedicated thread and the data between stages passes thru thread-safe FIFO queues.

Each processing thread waits in suspended state until any data appears in input queue. When new data is available in input queue the stage thread is resumed and starts its execution. The output data is passed to output queue and is available for next stage. The same approach is used for passing error reports in entire system.

8   System Performance Evaluation

The system has been evaluated experimentally with the use of two different platforms: desktop-class computer with modern CPU with 4 cores running at

3.4 GHz (i7-4770) and low power embedded system based on Raspberry Pi 3 platform (BCM2837). On both platforms 3 different comparison window sizes have been examined (one frame consists of 1152 coded samples - MPEG-1 Layer 2, 42 frames ∼ 1 s, 84 frames ∼ 2 s and 168 frames ∼ 4 s). To obtain stage processing time, statistics were gathered over 10 min during system operation and averaged. Test results have been summarized in Tables 1 and 2. Time required for data synchronization and packets queuing is very short and does not introduce a significant delay in system performance.

Table 1. Decoding time

Platform	Time required to decode one audio frame [ms] 48 kHz, 16 bit, 2 channels (stereo)
BCM2837	0.138
i7-4770	0.010

Table 2. Comparison time

Platform	Time of one comparison [milliseconds]		
	42 frames buffer (1 s)	84 frames buffer (2 s)	168 frames buffer (4 s)
Reference + one stream for comparison			
BCM2837	178.469	423.153	864.093
i7-4770	4.513	9.658	20.954
Reference + two streams for comparison			
BCM2837	184.949	477.901	875.243
i7-4770	4.973	10.780	21.279
Reference + four streams for comparison			
BCM2837	214.943	508.724	1171.427
i7-4770	4.981	11.011	21.384

9 Conclusions

The performance of the presented system for automated broadcast supervision proves to be effective in a real life application, even if the system is deployed on a low cost, low performance host like a single board computer. The system is able to process data faster than in real time, even if it has to process more than one audio stream. It can be used to supervise a set of programmes at once, still performing a real time supervision.

The system is able to detect different errors in audio stream assignment, like mismatch of audio streams, lack of audio, difference of audio levels exceeding a preset threshold as well as delay between audio streams exceeding a preset value. Small delays are inherently connected with the processes performed at the headend, so the system needs to make sure the delays are kept at normal level. The performance of the system is owed to the quality of the algorithm used for audio comparison and for its clear and efficient architecture for interfacing with the input data streams and communicating the current state of the system and signal errors to the outside world.

Acknowledgment. The presented work has been partially funded by the Polish Ministry of Science and Higher Education for the status activity consisting of research and development and associated tasks supporting development of young scientists and doctoral students in 2018 in Chair of Multimedia Telecommunications and Microelectronics.

References

1. ISO/IEC 13818-1:2000: Information technology – Generic coding of moving pictures and associated audio information: Systems
2. ISO/IEC 11172-3:1993: Information Technology – Coding of moving pictures and associated audio for digital storage media up to about 1,5 Mbit/s – Part 3: Audio
3. https://www.raspberrypi.org/products/raspberry-pi-3-model-b/ Accessed 27 May 2018
4. http://ffmpeg.org/about.html Accessed 27 May 2018
5. Lonvick, C.: The BSD Syslog protocol. RFC 3164 (2001). http://www.ietf.org/rfc/rfc3164.txt Accessed 27 May 2018
6. Lorkiewicz, M., Stankowski, J., Klimaszewski, K.: Algorithm for real-time comparison of audio streams for broadcast supervision. In: 25th International Conference on Systems, Signals and Image Processing, Maribor, Slovenia, 20–22 June 2018

Author Index

© Springer Nature Switzerland AG 2019
M. Choraś and R. S. Choraś (Eds.): IP&C 2018, AISC 892, pp. 253–254, 2019.
https://doi.org/10.1007/978-3-030-03658-4

Printed in the United States
By Bookmasters